Communication in the
CHIROPTERA

ANIMAL COMMUNICATION

EDITORS

Jean Umiker-Sebeok
Thomas A. Sebeok

PREVIOUSLY PUBLISHED

*Communication and Expression
in Hoofed Mammals*
by Fritz R. Walther

*Communication and Noncommunication
by Cephalopods*
by M. Moynihan

Communication in the
CHIROPTERA

M. Brock Fenton

INDIANA UNIVERSITY PRESS • *Bloomington*

Library of Congress Cataloging in Publication Data

Fenton, M. Brock (Melville Brockett), 1943–
    Communication in the Chiroptera.

    (Animal communication)
    Bibliography: p.
    Includes index.
    1. Bats—Behavior.   2. Bats—Physiology.
3. Animal communication.   I. Title.   II. Series.
QL737.C5F45   1985   599.4'0459   84-47965
ISBN 0–253–31381–3

1 2 3 4 5 89 88 87 86 85

CONTENTS

FIGURES

TABLES

ACKNOWLEDGMENTS

It has been my privilege to have interacted with many students on projects involving bats and/or communication. The list is long and includes, but is not limited to: Robert M. R. Barclay, Gary P. Bell, Jackie J. Belwood, R. Mark Brigham, David K. Cairns, Vanda Cuccaro, James H. Fullard, Caren Furlonger, Connie L. Gaudet, Judith F. Geggie, T. Michael Harrison, Marty L. Leonard, Robert M. Herd, Lynda S. Maltby, Alex M. Mills, Donald W. Thomas, Christine E. Thomson, and Tracey K. Werner. In addition I have benefited from discussions and interactions with many colleagues, including Jack W. Bradbury, Patricia E. Brown, David H. M. Cumming, Edwin Gould, Donald R. Griffin, Thomas H. Kunz, Gary F. McCracken, H. Gray Merriam, Gerhard Neuweiler, Paul A. Racey, James A. Simmons, Roderick A. Suthers, and Donald W. Thomas. I am particularly grateful to Donald W. Thomas and Roderick A. Suthers for reading the entire manuscript and making many helpful suggestions.

The assistance of Paul Chippendale, Vanda Cuccaro, and Theresa Vergette in matters of figures and literature cited made the task of preparing the manuscript much easier. I thank Gary F. McCracken for permitting me to use one of his photographs, and the University of Toronto Press for permission to reproduce a number of figures from *Just Bats*.

My studies of bat communication were greatly stimulated by spending part of a sabbatical leave from Carleton University at the Rockefeller University Center for Field Research and I again thank Drs. D. R. Griffin and P. Marler and other colleagues, including J. R. Baylis, R. J. Dooling, and D. E. Kroodsma, for such a rewarding stay there. The appearance of the American Society of Mammalogists' *Advances in the Study of Mammalian Behavior*, edited by J. F. Eisenberg and D. G. Kleiman, was timely in the context of the preparation of this manuscript since many of the ideas expressed therein were stimulating and relevant to this manuscript. My research has been generously supported by grants from the Natural Sciences and Engineering Research Council of Canada and by the Faculty of Graduate Studies and Research at Carleton University.

I will close by thanking my wife, Eleanor, whose tolerance and support have greatly enhanced my studies of bats.

Communication in the
CHIROPTERA

I

Introduction

Among the Mammalia, bats offer some of the best opportunities for studies of communication. A number of their characteristics make them ideal subjects for field or laboratory studies, and the array of dietary, social, and roosting situations which they exhibit provides a good background for experiments. The following scenarios will introduce the subject of bat communication.

Just before dawn, male *Saccopteryx bilineata* (Emballonuridae) return to their day roosts, which are located on the trunks of trees, check the boundaries of their roosting territories, and from there produce long songs directed at females returning from foraging. Between displaying to females, males approach other males trespassing on their territory, sometimes moving in parallel with neighbors along the boundaries of adjacent territories, vocalizing and shaking secretions from their wing glands on the substrate and on each other. Males joined by females in their roosting territories produce short songs and may mark the females with the secretions of the wing glands. Female-female interactions strongly influence the composition of the harems in roosting territories, although females do not produce long or short songs, or scent-mark other bats or the substrate. For more details see p. 84 and Bradbury and Emmons 1974.

At dusk as female *Nycticeius humeralis* (Vespertilionidae) prepare to depart from the day roosts housing their nurseries, they mark the faces of their young with exudate from the submaxillary glands. Later, when they return to nurse their young, females seem to rely on a combination of spatial memory, vocalizations of the young, and the scent marks to identify their own young, at least during the first phase of the infant's development (Watkins and Shump 1981). For more details about these interactions, see page 99.

Myotis lucifugus (Vespertilionidae) commonly forage over water or along the margins of streams and lakes, feeding in clumps of emerging insects, such as mayflies, caddis flies, or midges. This species locates its prey by echolocation, producing pulses of high-intensity sounds which sweep from about 80 to 40 kHz. Attempts to capture insects are accompanied by drastic increases in the rates of production of echolocation calls, known as "feeding buzzes." The calls one *M. lucifugus* uses to gather information about its environment through echolocation are used by conspecifics to find localized resources, such as patches of food, day roosts, mating sites, and hibernacula (Barclay 1982a). For more details about this situation, see page 66.

These three scenarios provide a glimpse of some of the signals and patterns of behavior involved in communication between bats. Typical of many other mammals, bats use some combination of auditory, visual, and olfactory signals to communicate information to others, or to gather information about the behavior of others. The importance of echolocation in the lives of some, but not all, bats has led workers to identify vocalizations as a prime channel for communication by bats, and in some species this is true. Bats, however, are also well equipped with a battery of glands which serve them in many roles when they interact with conspecifics, but which have been relatively little studied. And although the acuity of bats' echolocation, where it occurs, has given credence to the saying "as blind as a bat," many bats see very well, particularly under conditions of poor lighting. It is difficult to separate and identify specific roles for visual signals which are integral parts of many of the displays described for bats.

The tendency among students of bats to treat echolocation signals as entities separate from social calls has been widespread, but the distinction between the two kinds of signals is arbitrary. Signals in the behavior of mammals may simultaneously serve several functions and it is not always easy to identify specifically the "original" function of the display under study. Labelling the echolocation calls of bats—some bats or all bats—as displays important in communication may be inappropriate. The experimental evidence showing a communication role for echolocation calls falls well short of demonstrating that the animal producing the calls intended them to communicate information about itself or its activities.

In short, bats offer many excellent opportunities for students of animal communication. Included in the order Chiroptera are about 850 species which show a considerable diversity of diet, roosting habits, and social interactions. The diversity of bats in terms of number of species or the ecological backgrounds against which they operate offers many opportunities for experiments and investigations into the role of the environment in influencing the communication systems of these animals. (Fig. 1.)

The purpose of this book is to review the available information about communication in the order Chiroptera. The book is divided into five chapters. The first chapter is a brief review of aspects of the biology of bats which impinge directly on their communication behavior. Chapter 2 considers the apparatus bats have for sending and receiving signals in the visual, olfactory, and auditory domains. The third chapter is a treatment of the communication behavior of bats in a range of situations from feeding and roosting to mating and mother-young interactions. Chapter 4 presents case studies of four representative species of bats, and Chapter 5 is an overview of communication in the Chiroptera, including an effort to put the available information into a more general framework of communication. Readers familiar with bats might find it useful to proceed to the second or third chapter.

Bats have a reputation for being the most gregarious of mammals, and in some situations this reflects a high degree of sociality and sophisticated communication. In other instances, however, gregariousness is a by-product of limited roosting situations, and in many bat colonies there is little evidence of a strong underlying social structure. Despite their image as social and gregarious species, many bats are solitary, interacting with conspecifics only during mating and the rearing of young. Whatever their social bent, bats are among the most successful of mammals, reaching their highest taxonomic diversity in the tropics.

The last few years have seen a proliferation of information about bats, reflecting both an increased interest in them as subjects for study and technological advances that have opened new doors for working with them. Included in the array of information are several multi-

1–A

1–B

FIG. 1. The variations in the faces of bats reflect their diversity. Shown here are portraits of four species, two Megachiroptera (A, B) and two Microchiroptera (C, D). The African *Epomophorus gambianus* (A) and the New Guinean *Paranyctimene raptor* (B) are frugivorous, while the African *Scotophilus leucogaster* (C) and the African molossid *Tadarida ansorgei* (D) are insectivorous. Megachiropterans typically have simple ears, claws on their second

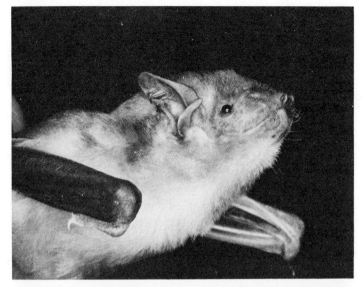

1–C

1–D

fingers, and no facial ornamentation; the function of the tubular nostrils of *P. raptor* remains unclear. The tragus, the prominent projection from the ear of *S. leucogaster*, is typical of some microchiropterans and is associated with echolocation, but in molossids this feature is not prominent. The faces of other bats are shown in Figs. 7, 8, and 11. Reproduced with permission of the University of Toronto Press.

author volumes dealing with different aspects of their biology (e.g., Wimsatt 1970a; 1970b; 1977; Slaughter and Walton 1970; Baker, Jones, and Carter 1976; 1977; 1979; Busnel and Fish 1980; Wilson and Gardner 1980; Kunz 1982b; Kunz in press). Several other books about bats are available (e.g., Allen 1939; Brosset 1966; Leen and Novick 1969; Barbour and Davis 1969; Yalden and Morris 1975; Fenton 1983a; Hill and Smith 1984), and papers about bats occur in a range of journals, reflecting ongoing work on many fronts. The network of biologists interested in bats is served by several specific newsletters: *Bat Research News, Myotis,* and *Australian Bat Research News.*

Given the number of publications, it is easy to forget that our knowledge about most species of bats is limited. This is particularly true in the area of communication. Relative to the overall literature, there have been few studies of bat behavior, partly because bats are nocturnal, live in hard-to-reach situations, and were assumed not to fare particularly well in captivity. Too many field studies of bats focused on their capture and on assessment of data on distribution and systematics while ignoring the dynamics of their behavior. Notwithstanding Griffin's early efforts to interest mammalogists in apparatus for eavesdropping on the echolocation calls of bats, relatively few workers exploited this convenient window on their behavior.

Several significant developments have altered this scene. One was, ironically, the development of an effective and convenient trap (Fig. 2) for catching many species of bats (Tuttle 1974). Another, the development of light tags, made it easier to observe their behavior in the field (Buchler 1976), an advance complemented by further miniaturization of radio transmitters (Fig. 3), permitting telemetry to be applied to several species of bats (e.g., Bradbury et al. 1979; D. W. Thomas 1979). These developments have coincided with the appearance of portable apparatus for studying the vocalizations of bats (Simmons, Fenton, Ferguson et al. 1979) and the realization that some species not only thrive in captivity, but can be trained to perform a variety of tasks (e.g., Simmons 1971; Rasweiler 1977; Tuttle 1982; Gaudet 1982; Gaudet and Fenton 1984). At the same time, apparatus for making observations at low light levels have opened our eyes to the behavior of bats under natural or captive situations. Furthermore, work on echolocation, notably with species capable of Doppler-shift compensation, has made some species of bats excellent subjects for research into information processing by mammalian brains. Another important contribution has

FIG. 2. The double frame harp or Tuttle trap is an excellent means of catching many species of bats. The bats fly into vertical strings (often 6 pound test monofilament fishing line) and slide down into the canvas bag. The narrow width of the bat and polyethelene flaps prevent the bats from flying out. Reproduced with permission of the University of Toronto Press.

Fig. 3. This *Euderma maculatum,* a 19 gram vespertilionid, is car-
rying a 0.9 gram radio transmitter with a range of about 1 kilo-
meter and a battery life of 10 days. Reproduced with permission of
the University of Toronto Press.

been putting the behavior of bats into a broader ethological context
(e.g., Bradbury 1977a).

It is interesting to note the time lags that have occurred in the study
of bats. Griffin (1958) made it easy for anyone to appreciate the potential for echolocation as a tool for researchers studying bats. Williams,

Williams, and Griffin (1966) used radio-tracking to study navigation in the phyllostomid *Phyllostomus hastatus,* and Griffin, Webster, and Michael (1960) used captive vespertilionids, *Myotis lucifugus,* to explore the use of echolocation in the location and capture of flying insects. In spite of these important contributions, there was considerable inertia in exploiting the technological developments involved, and workers have been slow to explore situations set up by earlier research (Griffin 1980).

In short, a combination of technological breakthroughs and more and more enthusiastic and prolific researchers has increased our knowledge of the biology of bats and focused attention on these animals. This setting makes it a propitious time to write about communication in bats, although it will quickly become clear that in spite of many published papers and considerable effort, our knowledge of communication in these mammals is in its infancy.

Bats constitute the second largest order in the class Mammalia, and although they are mainly tropical in distribution, they occur to the tree lines both north and south of the equator. Bats range in size from 2 grams as adults (the craseonycterid *Craseonycteris thonglongyai*) to over 1 kilogram (the pteropodid *Pteropus giganteus*). Most bats, however, are relatively small, and most species weigh less than 100 grams as adults (Fig. 4). The living species are arrayed in two suborders, the Megachiroptera and the Microchiroptera. Important differences between these two groups include the presence or absence of a claw on the second finger, the degree of specialization of the dentition and the pectoral girdle, and the use of echolocation as a means of orientation. Only some species occur in temperate regions, and these are usually members of the families Vespertilionidae and Rhinolophidae, animals capable of entering torpor and surviving prolonged inclement periods associated with temperate winters.

Their ability to fly makes bats extremely mobile, and their occurrence on remote oceanic islands, including Hawaii, underscores their potential for dispersal. Some species make considerable annual migrations, moving to and from areas where food is seasonally abundant. Despite the general acceptance of these patterns of movement, detailed knowledge about the migrations of most bats is lacking and only extensive and intensive banding studies in temperate zones have given a clear picture of the magnitude of some seasonal movements (Griffin

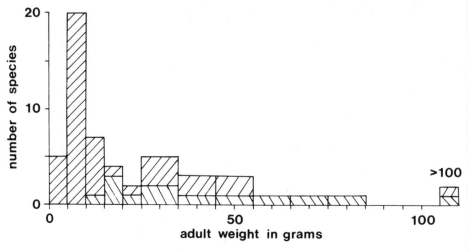

FIG. 4. This histogram emphasizes the fact that bats tend to be small (in weight) mammals. Arranged here are the Chiroptera which occur in Malaya (from Fenton and Fleming 1976), showing the number of species in different weight classes and distinguishing between the Megachiroptera \\\\\ and the Microchiroptera /////. A similar plot of a Neotropical fauna would emphasize this feature of bats even more, as even fewer species there exceed 100 grams adult weight.

1970). The migrations of tropical bats are receiving more attention (D. W. Thomas 1983), but the data base is still limited (Fenton and Thomas in press).

Bats have a reputation for being the most social of mammals; some species are certainly the most gregarious, but, as we shall see, aggregations of bats often reflect localized resources rather than underlying social structures. While there are species that occur at very high densities in their roosts, for example the molossid *Tadarida brasiliensis,* which forms nursery colonies composed of millions of individuals, many others (e.g., the vespertilionid *Lasiurus borealis*) are solitary except when rearing current young.

Coincident with their notoriety as gregarious mammals is a reputation for being dirty. Although the bats studied to date spend great amounts of time each day grooming themselves, they harbor a wide range of ectoparasites. Social or gregarious species tend to have a higher diversity of ectoparasites than solitary forms. A detailed consideration of the lives of some bat ectoparasites is reviewed by Marshall (1981; 1982). The relationships between population density and endoparasites are less clear (e.g., Ubelaker 1970), but some forms transmitted by ectoparasites, such as blood trypanosomes, may be expected at higher incidence among social or gregarious species (Bower and Woo 1981a; 1981b).

EVOLUTION AND CLASSIFICATION

Icaronycteris index is the first known fossil bat, a well-preserved specimen from the lower Eocene of Wyoming (Jepsen 1970), a time when bats were easily recognizable as bats and whales were not yet so specialized (Gingerich et al. 1983). Although it has a claw on its second finger, normally a trait of the Megachiroptera, *Icaronycteris index* is clearly microchiropteran in its affinities and provides no indication of which group of mammals is most closely linked to the evolution of bats. The fossil record of bats is not particularly good, although many currently recognized families are known from Oligocene deposits in different parts of the world (Jepsen 1970; Koopman and Jones 1970; Smith 1976; Sigé 1977; Archer 1978; Sigé and Russell 1980; Table 1). Smith (1972) suggested that a paleochiropteran grade of bats had appeared and radiated by the Eocene, when it was widespread and gave rise to the current Microchiroptera. Unfortunately, the fossil record does little to help us to understand the relationships between different major units of bat classification. It is generally agreed that the Chiroptera and the Insectivora share a common ancestor (e.g., Eisenberg 1981).

There is some question among biologists about whether the Megachiroptera and the Microchiroptera are more closely related to one another than either is to any other group of mammals; in other words, is the order Chiroptera monophyletic? Despite the general similarity, particularly in wing structure, there are important anatomical differences which may suggest a remote relationship between the two groups. The history of the situation is presented by Smith (1980). Summaries and comparisons are provided by several workers (e.g., J. K. Jones and Genoways 1970; Smith 1976; 1977; 1980; Luckett 1980; Koopman and MacIntyre 1980; Bhatnagar 1980; Novacek 1980; Suthers and Braford 1980; Smith and Madkour 1980). One classification of the Chiroptera (Fig. 5) illustrates possible interrelationships of different groups. This arrangement, however, is not uniformly accepted, and Van Valen (1979), for one, proposed some radical departures from it. Ongoing work may produce further changes. For example, Pierson et al. (1982) used a range of features, including anatomy

TABLE 1. Classification, fossil record, distribution, diet, and diversity of the Chiroptera (data from Koopman and Jones 1970 and Smith 1976)

FAMILY	FIRST FOSSILS		DISTRIBUTION	DIET	NUMBER OF EXTANT SPECIES
	age	*name*			
Palaeochiropterygidae	Eocene	*Icaronycteris*	North America	—	0
	Eocene	*Paleochiropteryx*	Europe	—	0
	Eocene	*Archaeonycteris*	Europe	—	0
	Eocene	*Cecilionycteris*	Europe	—	0
	Eocene	*Ageina*	Europe	—	0
	Eocene	*Matthesia*	Europe	—	0
Pteropodidae	Oligocene	*Archaeopteropus*[1]	fossil from Europe; now Old World tropics.	fruit and nectar	150
Rhinopomatidae	—	—	North Africa to southern Asia and Borneo	insects	2
Craseonycteridae	—	—	Thailand	insects	1
Emballonuridae	Eocene	*Vespertiliavus*	fossil from Europe; now pantropical	insects	44
Megadermatidae	Eocene	*Necromantis*	fossil from Europe; now Africa, India, East Indies and Australia	insects to vertebrates	5
Nycteridae	—	—	Africa, Madagascar, Malaysia, Java, Summatra	insects to vertebrates	13
Rhinolophidae	Eocene Eocene	*Paraphyllophora* *Rhinolophus*	Old World	insects	69

Family	Epoch	Genus	Distribution	Diet	No.
Hipposideridae	Eocene	*Hipposideros*	Old World tropics	insects	56
Phyllostomidae	Oligocene	*Notomycteris*	New World tropics	fruit, nectar, pollen, insects, vertebrates, blood	123
Mormoopidae	—	—	New World tropics	insects	8
Noctilionidae	—	—	New World tropics	insects and fish	2
Natalidae	—	—	New World tropics	insects	4
Furipteridae	—	—	New World tropics	insects	2
Thyropteridae	—	—	New World tropics	insects	2
Vespertilionidae	Eocene	*Stehlina*	fossils from Europe; cosmopolitan	insects to fish	283
	Eocene	*Nycterobius*			
Mystacinidae	—	—	New Zealand	insects to fruit	1
Myzopodidae	—	—	Madagascar	insects	1
Molossidae	—	—	pantropical	insects	82

[1] possibly not a Pteropodidae; Smith 1976.

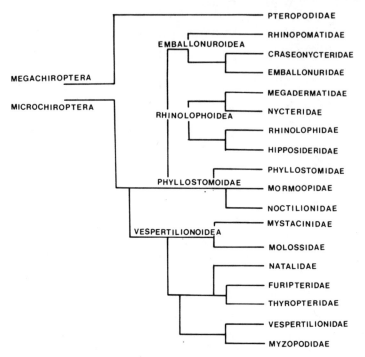

FIG. 5. This arrangement of the suborders (Megachiroptera and Microchiroptera), superfamilies (Emballonuroidea, Rhinolophoidea, Phyllostomoidea, and Vespertilionoidea), and families of living bats provides an indication of possible relationships between groups (modified from Eisenberg 1981).

and serum proteins, to include the endemic New Zealand Mystacinidae among the Phyllostomoidea. Although the vampire bats are now usually included as a subfamily of the Phyllostomidae, they have been treated as a separate family, the Desmodontidae. Evidence from serum proteins, karyotypes, and morphology makes it clear that the vampires represent an offshoot of the phyllostomids, and many biologists do not feel that the differences warrant treatment as a separate family (e.g., Forman, Baker, and Gerber 1968). On the other hand, Smith (1972) makes a convincing case for treating the Mormoopidae as a family separate from the Phyllostomidae, rather than as a subfamily (the Chilonycterinae). Neither view is unanimously accepted (e.g., Hall 1981), and some important questions remain about the relationships between different groups of bats.

In at least one species, *Uroderma bilobatum,* a phyllostomid, there are karyotypically distinct subspecies, a situation that has produced important genetic studies (e.g., Baker, Bleier, and Atchley 1975; Baker 1981), lamentably none yet supported by behavioral observations. In some cases hybrids between different species of bats have been proposed. For example, morphological studies of *Myotis lucifugus* and *Myotis yumanensis,* vespertilionids from western North America, indicated intermediate forms, possibly hybrids (Harris 1974; Parkinson 1979). Work in western Canada revealed additional morphological intermediates between these two species, but the specimens in question were genetically distinct, either one species or the other. Furthermore, field observations showed that the two species had different patterns of habitat use and associated differences in diet (Herd and Fenton 1983).

Differences in behavior have influenced decisions about classification. Lawrence and Novick (1963) used behavioral differences to support treating *Lissonycteris* as a genus separate from *Rousettus.* More recently, McWilliam (1982) used differences in mating patterns and other evidence to distinguish *Hipposideros commersoni* from *Hipposideros gigas* in east Africa; these two forms had been considered subspecies (Rosevear 1965).

The differences between the Megachiroptera and the Microchiroptera are striking, and are partly associated with life styles (e.g., Koopman and Jones 1970; J. K. Jones and Genoways 1970). Cranially, the Megachiroptera, particularly in their dentition, reflect specializations for dealing with fruit and nectar. Postcranially, however, the Megachiroptera are relatively unspecialized, reflecting their laborious flight, which serves mainly to get them from one place to another. In comparison, most Microchiroptera have less specialized teeth and are more specialized postcranially. The Microchiroptera with the most specialized teeth have the most specialized diet—blood. The greatest similarity in dentition between the two suborders occurs in some frugivorous phyllostomids (G. S. Miller 1907). Postcranial specializations of microchiropterans, particularly of the shoulder girdle, reflect the fact that many of them forage on the wing.

Another main difference between the two suborders is the use of echolocation. The Megachiroptera, with the known exception of some species in the genus *Rousettus,* do not echolocate, while all of the Microchiroptera studied to date (albeit, a minority of species) can use echolocation. Furthermore, echolocating *Rousettus* spp. produce their orientation sounds by clicking their tongues, while in the Microchiroptera for which there are data these sounds are produced in the larynx.

FLIGHT

The ability to fly makes bats unique among mammals and provides them with their most distinctive features, their wings. In addition to serving in flight, the wings of bats are used to capture flying insects, perhaps to counter last-ditch evasive maneuvers by insects (Humphries and Driver 1970). Wings also serve in temperature regulation, as fans or as surfaces from which heat effectively can be lost. Heat loss from the wings is facilitated by direct anastomoses between small arteries and veins in the wings of some bats (Kallen 1977). Wings are also used in a variety of behavioral displays, from immobilization of females during copulation to attraction of females to mating sites. Recent reviews of the anatomical specializations of bats for flight are provided by several publications (e.g., Vaughan 1970a; 1970b; 1970c; 1970d; Norberg 1969; 1976a; 1976b; 1976c; Strickler 1978; Altenbach 1979; Hermanson and Altenbach 1983).

The wings of bats are folds of skin stretched between elongated finger bones. The forearm and humerus and long bones of the hind limb are also elongated. In flight the wing area proximal to the fifth digit provides lift, while the portion distal to the fifth digit acts as a variable pitch propeller providing thrust. Power comes from the contraction of nine pairs of flight muscles located on the chest and back (unlike the arrangement of wing elevators and depressors in birds). Other specializations for flight include the structure of the flight muscles (Armstrong, Ianuzzo, and Kunz 1977) and the ability to maintain a taut wing membrane, in some species achieved by a sheet of muscle in the wing membrane itself (Gupta 1967). Recently hair sensors on the surface of the wing have been proposed as a means of measuring airflow and providing feedback permitting the bat to adjust tension on the wing membranes (Fowler and Zook 1982). Edgerton, Spangle, and Baker (1966) reported that *Tadarida brasiliensis* altered the shape of their wings from narrow to broad depending upon the situation in which they were operating.

The leading edge of the wing influences patterns of airflow and maneuverability in flight (Norberg 1969). In some species the scapula rocks about its long axis, providing an extra functional joint to the wing. The upstroke in some species is stopped by a locking arrangement of the humerus against the scapula, a doorstop feature that is best developed in molossids, the bats usually considered the most highly adapted for rapid flight (e.g., Vaughan 1970c).

Studies of energy consumption, albeit to date conducted with larger (c. 60 g) bats, indicate energetic costs of flight comparable to those reported for birds of similar size (S. P. Thomas 1975). Differences in wing shape, wing area, and wing tip among the microchiropterans result in striking differences in flight ability. Broadwinged species show highly maneuverable flight, perhaps best seen among the megadermatids and nycterids, but also common in many phyllostomids. Species with longer, narrower wings, notably molossids, have less maneuverable flight, and in some cases their roosts must offer an initial free drop of several meters to allow the bats to gain enough airspeed to assume horizontal flight (Vaughan 1959).

In comparison with birds, bats have two conditions making them potentially less efficient flyers (Jepsen 1970; Feduccia 1980). The first of these is bearing live young, which burdens females particularly during the latter stages of pregnancy. Another disadvantage is teeth, relatively dense structures at the front end of the body. The degree of handicap presented by either condition is difficult to assess. Some females carry their young when they go to feed, and others, notably some of the carnivorous species can carry loads of almost 75 percent of their mass in their mouths as they fly to their roosts.

ANATOMY

The bodies of bats are basically mammalian, and most species have fur, the exception being naked bats (*Cheiromeles* spp.; Molossidae) from southeast Asia. As noted above, a claw is absent from the second finger of living Microchiroptera, but the Megachiroptera, with the exceptions of some species (genera, *Dobsonia, Eonycteris, Melonycteris,* and *Notopteris*), have claws on their second fingers. The Furipteridae excepted, all bats have conspicuous thumbs with claws. In some species the thumbs are conspicuously long, and in *Desmodus rotundus,* the common vampire, this is a modification for taking off from the ground, with the elongated thumb acting like a throwing stick to help launch the bat (Altenbach 1979).

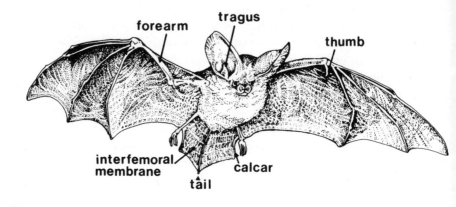

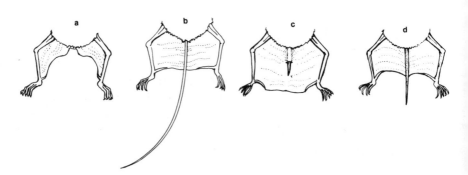

FIG. 6. This illustration provides a basic map to a bat, showing some
of the diagnostic features and variations in the arrangement of the
tail which occur in the Chiroptera. Some bats lack an external tail
(a) and have interfemoral membranes modified accordingly, a
situation common among the Phyllostomidae and the Pteropodidae. In the Rhinopomatidae the tail is long and mouse-like (b), a
situation presumed to have been typical of some of the early bats.
In the Emballonuridae the tail protrudes through the interfemoral
membrane (c), while in the Molossidae the thick tail protrudes
from the end of this membrane (d). In most bats the tail is fully
enclosed in the interfemoral membrane, shown in the flying *Plecotus auritus,* a vespertilionid. Sketch by Connie L. Gaudet, reproduced with permission of the University of Toronto Press.

There is considerable variation in the tails of bats (Fig. 6). In the Rhinopomatidae the tail is very long and slender, extending well beyond the posterior margin of the interfemoral membrane. This condition is presumed to resemble that of the paleochiropteran grade (Smith 1976). In other species, including some pteropodids, phyllostomids, and the craseonycterid, there is no external tail. In other bats tails are long and entirely enclosed by the interfemoral membrane. In the Nycteridae, the tail terminates in a T-shaped cartilage; in the emballonurids it protrudes through the interfemoral membrane, and in molossids the thick tail extends well beyond the posterior margin of this membrane.

The skulls of bats vary in shape, reflecting the diversity of faces in the order. The teeth are also variable and heterodont (Slaughter 1970). Many species of bats have milk teeth, erupted at birth or which appear shortly after birth. The deciduous dentition is markedly different from the adult dentition and apparently allows young to cling to their mothers' teats. The deciduous dentition is heterodont and the incisors are most specialized in form; the canines and premolars are usually simple and peglike (Fenton 1970; Phillips 1971; Birney and Timm 1975).

The incisor teeth of bats vary considerably, from being prominent and robust to being tiny or absent. The upper incisors of vampires are specialized for making a superficial wound and permitting the bats to feed. At the other extreme are megadermatids, which lack upper incisors. Lower incisors are often reduced in nectar-feeding bats, and in many molossids they are situated between prominent cingula on the canines and may serve only in grooming.

Canines are prominent in most bats, notably vampires (Greenhall, Joermann, and Schmidt 1983), and premolars vary from minute to massive. In animal-eating bats, the molar teeth usually have V- or W-shaped cusps that provide a combined crushing and shearing action (Slaughter 1970). The molars of pteropodids, which are mainly frugivorous, tend to be broad and flattened, a condition approached by some phyllostomids (G. S. Miller 1907). In nectar and pollen feeders, the molars and premolars are often reduced in size. In *Desmodus rotundus,* shearing action between the upper canine and the lower premolars may be used to remove hair or feathers from a potential bite site (Greenhall 1972).

In contrast to the Megachiroptera, which have relatively simple,

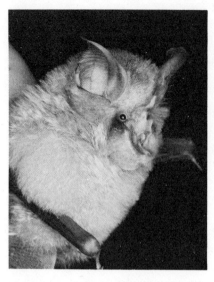

7–A

7–B

7–C

7–D

7–E FIG. 7. The facial ornamentation of some bats is presumed to be   7–F
influenced by (or to influence) their echolocation, although the
direct relationships are not at all clear. Shown here are three
species of rhinolophoids (*Rhinolophus simulator*—A; *Megaderma
lyra*—B; and *Nycteris grandis*—C), two phyllostomids (*Macrotus californicus*—D; and *Desmodus rotundus*—E), and a vespertilionid,
Myotis septentrionalis (F). All of these bats except the vespertilionid
probably emit their ultrasonic echolocation calls through their
nostrils; the heat-sensitive pits of the vampire *D. rotundus* are
located behind the nose leaf. Reproduced with permission of the
University of Toronto Press.

doglike faces, the faces of the Microchiroptera often bear bizaare-
looking ornaments. Some families are characterized by their facial
ornamentations, presumed to be involved with their use of echolocation. In some species there are prominent nose leafs; in others, wartlike
projections, papillae, deep slits, and/or active glands. The range of
structures is considerable and there is no reason to assume that all of
them serve the same function. The eyes of megachiropterans and
many microchiropterans are conspicuous; in other microchiropterans
they may be less so. The visual acuity of bats varies, but is less adversely
affected than man's by declining levels of light. (Figs. 1 and 7.)

In some species there is striking sexual dimorphism. In the pteropodid *Hypsignathus monstrosus,* males are larger than, and very distinct
from, females, and these differences are associated with the displays
males use to attract females. In other species the differences may be
equally marked, but the functional details remain obscure or un-

known. For example, in the phyllostomid *Ametrida centurio*, males are much smaller than females, but there are no behavioral data to put this difference into context (Peterson 1965a). In some African molossids, notably *Tadarida chapini*, males have conspicuous erectile interaural crests, structures that are reduced and inconspicuous in females and in subadult males (Fig. 8). Sexual dimorphism in the vespertilionid *Lasiurus borealis* involves differences in color; males are bright red and females a duller red (Barbour and Davis 1969). In many species of molossids there is marked sexual dimorphism in the premolar teeth (G. E. Turner 1970), which has not been explained in any functional way. Other manifestations of sexual dimorphism include glands, usually most conspicuous in males and sometimes lacking in females. In most bats males tend to be slightly smaller than females, a difference usually attributed to greater wingloading of females during the terminal stages of pregnancy (e.g., D. F. Williams and Findley 1979).

In most respects the anatomy of bats is essentially mammalian, although modifications for flight have affected every major organ system (e.g., Vaughan 1970a; 1970b; 1970c; 1970d).

ECHOLOCATION

Echolocation, a system of orientation involving animals' use of sounds they produce to detect objects in their path, was first described for bats, but is not characteristic of all bats (Griffin 1958). Among the Megachiroptera only some species in the genus *Rousettus* use echolocation; the others orient themselves by vision, relying to some extent on olfaction to detect and identify appropriate food sources. All of the Microchiroptera studied to date can echolocate, and some rely on echolocation to detect the insects on which they feed (Griffin, Webster, and Michael 1960; Simmons, Fenton, and O'Farrell 1979; Pye 1980). It is clear, however, that despite their considerable ability in the field of echolocation, many microchiropterans rely on other forms of orientation when hunting. For example, Fiedler (1979) showed that the megadermatid *Megaderma lyra* might or might not use echolocation when hunting mice, and Tuttle and Ryan (1981) convincingly demonstrated that the phyllostomid *Trachops cirrhosus* locates the frogs on which it feeds by listening to the songs of the males. Bell (1982a) found that the vespertilionid *Antrozous pallidus* used sounds emanating from moths and did not produce echolocation calls during its attacks, while the phyllosto-

FIG. 8. Male *Tadarida chapini* (A) have prominent interaural crests which are erectile, while these are poorly developed in females (B). In subadult males the crest is also little developed. In adult males the base of the crest is associated with a prominent gland, a feature common to some other molossids which lack this degree of development of the crest. *Tadarida chapini* weigh about 10 grams and are widespread but uncommon in Africa. Reproduced with permission of the University of Toronto Press.

mid *Macrotus californicus* used either vision or sounds emanating from its targets to detect them (Bell 1982b).

Not all bats echolocate and not all bats with the ability to echolocate use it to find their food. D. J. Howell (1974) showed decreasing acuity of echolocation among glossophagine phyllostomids increasingly specialized for exploiting nectar and pollen as food. The role of echolocation in the lives of frugivorous phyllostomids is not clear, since olfaction is obviously important in their food locating and evaluating activities (Fleming 1982; Heithaus 1982). Furthermore, not all bats take the same approach to echolocation. There are important differences in the features of their calls, reflecting differences in the information the bats are trying to obtain.

As noted earlier (Fig. 7), the facial ornamentation of bats may influence echolocation, perhaps by affecting the patterns of sound radiation from the bat (e.g., Griffin and Novick 1955; Novick 1958; Grummon and Novick 1963; Grinnell and Schnitzler 1977; Schnitzler and Grinnell 1977). The tragus, a conspicuous structure in the pinnae of many bats, functions in vertical localization of targets, according to work by Lawrence and Simmons (1982a) with the vespertilionid *Eptesicus fuscus.*

The echolocation calls of microchiropterans are distinctive, typically showing marked changes in frequency over time. These structured calls differ from clicks, which are brief, discrete, unstructured broadband signals with a rapid onset. Clicks, as Gould (1983) points out, are easy to produce, hear, and locate. Clicks are used by most other echolocating animals, including the echolocating species of *Rousettus* (Wood and Evans 1980; Fenton 1980). It is possible to characterize the components of calls used in echolocation by Microchiroptera (Fig. 9). In some species the calls are constant frequency (CF); in others the bandwidth is wider, as the calls typically sweep from high to low frequency. These frequency modulated (FM) calls may be steep, rapidly covering a wide range of frequencies, or shallow, calls of narrower bandwidth and longer duration. Although these terms—CF, FM, or some combination thereof—can be used to describe the vocalizations, they do not necessarily portray the strategies of the bats (Gustafson and Schnitzler 1979; Fenton 1982a).

All echolocating bats do not produce orientation sounds in the same way. Many echolocating bats emit their sounds through their open mouths, but some, notably the rhinolophoids and perhaps many of the

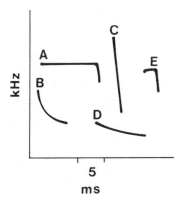

Fig. 9. A comparison of a variety of call designs from echolocating Microchiroptera to illustrate constant frequency (CF) with a terminal steep frequency modulated (FM) sweep—A; a combination of steep and shallow FM sweep—B; a broadband steep FM sweep—C; a shallow FM sweep—D; and a call with an initial rise in frequency followed by a brief CF component and ending in a steep FM sweep—E. Calls dominated by CF components (A) are associated with Doppler-shift specializations and are typical of species in the families Rhinolophidae and Hipposideridae, and in one species of Mormoopidae. There is considerable variation in bats producing other calls, although B and C are typical of many vespertilionids, and D of some vespertilionids and molossids. Some emballonurids produce echolocation calls similar to E. The vertical scale is frequency (kHz—kilohertz) and is unlabelled as these calls do not refer to specific bats; the horizontal scale is time in ms (milliseconds). Calls A, B, C, and E are often entirely ultrasonic; many species using calls as shown in D include lower-frequency components and may be quite audible to the unaided human ear.

phyllostomids, emit their echolocation sounds through their nostrils. Emission of sounds through the nostrils should allow the bat simultaneously to echolocate and chew.

Differences in the structure of echolocation calls associated with bandwidth influence the information the bats obtain through echolocation (Simmons and Stein 1980). CF calls with narrow bandwidths are excellent for detecting targets, and increasing the bandwidth, in some cases by the addition of harmonics or overtones, increases the precision with which the bats locate targets (Simmons and Stein 1980). An impressive display of this relationship is provided by the changes in call design as bats search for, detect, and close with flying targets (Fig. 10).

Errors associated with Doppler shift may be important for echolocating bats, particularly those relying on CF signals. Calls with broader bandwidths are less vulnerable to this source of error (Simmons and Stein 1980). Two species which should be vulnerable to Doppler-shifted errors because of their use of signals of narrow bandwidth are *Taphozous mauritianus* and *Rhinopoma hardwickei* (Fenton, Bell, and Thomas 1981; Habersetzer 1981, respectively). Some molossids hunting in open areas with little clutter, particularly at higher altitudes, use shallow FM calls and may operate in spite of some of the problems associated with these narrow-band signals because of the lack of clutter (Griffin and Thompson 1982). Simmons, Lavender, et al. (1978) found that the molossid *Tadarida brasiliensis* added harmonics to its calls when hunting in cluttered settings.

Bats in the families Rhinolophidae and Hipposideridae, and one species in the Mormoopidae, have been called "Doppler-shift compensators" because of neurological and anatomical specializations for exploiting Doppler-shifted echoes. These bats should be very effective at detecting flying targets (e.g., Schuller and Pollak 1979; Neuweiler 1980a; Schnitzler and Flieger 1983; Vogler and Neuweiler 1983; Bell and Fenton 1984).

Bats adjust the rates at which they produce vocalizations used in echolocation. Individuals cruising and searching for targets produce echolocation calls at low rates (10 to 50 sec^{-1}); when closing with flying insects the same individuals greatly increase their rates of pulse production (to over 500 sec^{-1}). High-pulse repetition rates associated with attacks on insects (Fig. 10) are called "feeding buzzes," an unfortunate label suggesting success rather than attempted capture. Increasing rates of pulse production may ensure that the bat tracks prey accurately as the range decreases, and some bats drop the feeding buzz when hunting stationary targets (e.g., the vespertilionid *Myotis auriculus;* Fenton and Bell 1979). The situation, however, is not clear, since some other species, such as the nycterids *Nycteris grandis* and *N. thebaica* (Fenton, Gaudet, and Leonard 1983), produce feeding buzzes as they close with targets even when they are relying on other cues to locate prey.

Echolocation by bats is a relatively short-range operation. The only published data on the threshold of detection (as opposed to the distance of reaction) are provided by the work of Kick (1982), who determined that an echolocating *Eptesicus fuscus* first detected a 19 mm diameter sphere at a range of 5 m. Observations of the hunting be-

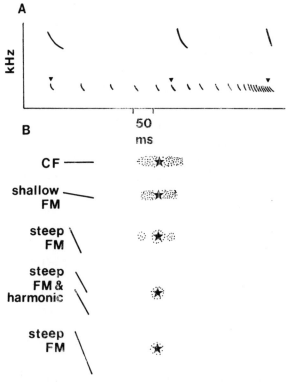

FIG. 10. This shows the sequence of calls associated with an attack on a flying insect by a vespertilionid (A), and the influence of design of echolocation call on the precision of target location by a bat (B). The high pulse-repetition rate at the end of the attack (A) is referred to as a "feeding buzz." Calls in the sequence marked ▼ are shown in more detail above to illustrate how call design changes through the sequence of attack. In B is a representation of how different call designs influence target resolution, following the suggestions of Simmons and Stein (1980). The target is shown as a star; the stippled areas represent the zone in which the bat would perceive it using the calls illustrated. The CF call is excellent for detecting a target but provides little precise information about its location. Increasing the bandwidth (= frequencies covered by) of the signals (shown as sonagrams) from shallow to steep FM increases the precision of target location. Further increasing the bandwidth by the addition of harmonics (overtones) or the production of a longer sweep permits more precise target location.

havior of bats reported in the literature suggest that some species react
to targets at 15 m (the emballonurid *Taphozous peli;* Brosset 1966), and
10 m is relatively common (e.g., the hipposiderid *Hipposideros commersoni,* Vaughan 1977; or *Taphozous mauritianus,* Fenton, Bell, and Thomas
1981). The field data, however, do not provide accurate measures of
the distance at which the bat reacted or about the size of the target(s)
involved. The field observations do not preclude the bats' responses to
other stimuli, such as sounds emanating from the prey.

Suthers (1970a) proposed that bats rely on vision to extend their
range of operation, and this is supported by work on *Eptesicus fuscus,*
which use visual and other acoustical cues when looking for hunting
grounds (Childs and Buchler 1981; Buchler and Childs 1981). Limited
range of operation makes it easier to understand why bats relying on
echolocation are relatively small. If their range of operation is put
into the context of their size, *E. fuscus* first detect spheres 19 mm in
diameter (16 percent of their total length) at a distance of 40 total
lengths (head, body, and tail), according to Kick's (1982) results.

Initial intensity of echolocation calls influences the range of operation, and bats relying on lower-intensity calls should have shorter
operational ranges than those using high-intensity calls. The relationships between intensity of echolocation calls and operational range
of echolocation remain, however, relatively unexplored. Atmospheric
attenuation of sound is not even across the frequency spectrum normally used by echolocating bats, and this further influences the operational range. Griffin (1971) and Lawrence and Simmons (1982b) have
shown that higher-frequency sounds are more rapidly attenuated than
lower-frequency sounds.

Frequency and intensity of bat echolocation calls also bear on a major
disadvantage of echolocation, the conspicuousness of echolocators.
Roeder (1967) showed that many moths have ears sensitive to the
high-frequency cries of echolocating bats, and the same is true of
lacewings (recent reviews Fenton and Fullard 1981; L. A. Miller 1982).
Roeder showed for some moths, and Miller for some lacewings, that
insects which hear bats have 40 percent less chance of being caught
than deaf insects. Furthermore, audiograms of a variety of moths show
how adjustments to the intensities and frequencies of echolocation calls
drastically alter the distances at which bats are detected by these insects
(Fenton and Fullard 1981; Fullard and Thomas 1981; Fullard 1982a;
Fullard, Fenton, and Furlonger 1983). It is easy to see how hearing in

insects could influence the design of bat echolocation calls. The crucial thing for the bat probably is altering the design of its call to achieve detection of the insect target before the bat's call is, in turn, detected.

There is a huge volume of literature on the details of echolocation (Fig. 1 in Grinnell 1980; Busnel and Fish 1980). Echolocation works because each outgoing call is registered in the bat's nervous system and is then compared to its returning echo; the difference between what went out and what comes back is the information used by the bat (in addition to time cues). This means that bats must avoid deafening themselves (e.g., Henson 1970a; Jen and Suga 1976) and that to exploit the echoes of calls of another individual would be very difficult if not impossible. If this view of the operation of echolocation is correct, jamming and interference by different individuals should be minimal. Changes in the frequencies of echolocation calls coincident with the behavior of other bats have been interpreted as jamming avoidance (e.g., Habersetzer 1981), although some bats are very resistant to jamming (Griffin 1958). The jamming function hypothesized for the clicks of some arctiid moths (Fullard, Fenton, and Simmons 1979) is interference with information processing by the bat, not mimicking of the echoes expected by the bat.

Echolocation makes bats conspicuous, and the vocalizations associated with echolocation are obviously available in other contexts. The calls bats use in echolocation appear to have precursors in form' and function in communication (e.g., Gould 1977a). Although echolocation is commonly associated with bats, it is not a uniform trait, either in its incidence, in the conditions of its use, or in the details of its operation.

Diet and Feeding Behavior

The diets of bats reflect their taxonomic diversity. Although insectivorous forms account for about 70 percent of the known species, around 20 percent regularly take fruit and 8 percent feed on nectar and pollen. In addition, some species increase the range of food consumed; these include three species of blood-feeding vampires, several fish eaters, and a number of carnivorous species that include a range of terrestrial vertebrates, such as frogs, birds, mice, lizards, and other bats, in their diets. There is enough variation in the diets of some bats to obliterate the distinction between insectivorous and carnivorous, as, for example, in the African *Nycteris grandis*. This 35 to 40 g bat commonly eats insects

ranging in size from small caddis flies to large (10 g) dung beetles, and
vertebrates ranging from fish and frogs to birds and bats (Fenton,
Thomas, and Sasseen 1981). In reviewing the diets of bats, I consider
first species relying mainly on animals and distinguish them from
others feeding mainly on plant products. Even this distinction, however, is not clear, as some insectivorous species often eat fruit (e.g.,
Daniel 1979; D. J. Howell 1980) and others feeding mainly on plant
products also include insects in their diets (e.g., Ayala and D'Alessandro 1973; Gardner 1977).

Animal Protein

With the exception of the vampires, which feed only on blood, most
bats that feed on animal protein rely on insects to some extent. Larger
species, particularly those taking prey from the ground or foliage
(gleaners), often include larger prey items in their diets, from arthro-
pods to vertebrates. There is a record of the 15 to 20 g *Antrozous pallidus*
taking a pocket mouse (*Perognathus* spp; Bell 1982b). The diet of any
species of bat should reflect its size (e.g., Anthony and Kunz 1977),
although gleaners appear to regularly take proportionally larger prey
than do aerial-feeding bats. Larger species can include larger prey
items in their diets, so that a 40 g *Trachops cirrhosus* has access to a
greater spectrum of prey than the 2 g *Craseonycteris thonglongyai*. Available evidence suggests that some of the larger bats often take smaller
insects (Fenton, Thomas, and Sasseen 1981).

Animal-eating bats use several foraging strategies. Some species
hunt flying prey, others take nonvolant prey. Those pursuing flying
targets may fly continuously while hunting, while others wait on a perch
for targets to fly within range. Species taking nonflying prey also show
these two approaches. Among flying species chasing flying prey, some
(usually smaller) detect and react to targets at short range (1–2 m) while
others (usually larger) operate at longer range, perhaps to 15 m. This
difference in reaction distance influences feeding potential. Bats specialized for operation at short range maneuver within a swarm of flying
insects, making repeated attempts to capture food, while species operating at longer range usually make only one capture attempt per pass
through the patch (Fenton 1982a).

Some species of bats may be relatively restricted to a particular
feeding strategy. For example, the vespertilionid *Myotis lucifugus* typically flies continuously while hunting, pursuing flying prey most of the

time (Fenton and Barclay 1980). These bats, however, occasionally glean insects from the surface of the water (Fenton and Bell 1979; Harrison 1983), and young have been observed waiting on perches for passing prey (Buchler 1980a). Other species are more flexible. *Cardioderma cor*, an African megadermatid, shows diversity in foraging behavior and prey consumed. In some settings these bats make short flights from perches to attack prey on the ground or, occasionally, flying prey. They may also fly continuously in pursuit of flying prey, sometimes smaller bats (Vaughan 1976). Unfortunately, we lack details about the foraging behavior of most animal-eating bats.

I think it is premature to identify particular bats with any one feeding strategy (Fenton 1982a), in spite of the observed trends. An attribute common to many animal-eating bats is opportunism, conspicuous when bats feed among concentrations of insects such as those aggregated at lights (Fenton and Morris 1976). Exploitation of clumped resources reflects flexible foraging strategies and adaptations for rapid mastication of prey (Kallen and Gans 1972).

Flexibility in foraging behavior and opportunism are reflected in the variable diets typical of most species studied to date (reviewed in Fenton 1982a). There are, however, occasional examples of species specializing on a particular group of insects. The best one to date is the African hipposiderid *Cloeotis percivali*, which feeds heavily on moths (Whitaker and Black 1976). In many other cases, detailed studies indicate that species characterized as specialists on a particular group of insects are not. One example is provided by the vespertilionid *Lasiurus cinereus*, identified by Black (1972) as a moth specialist. These bats may feed on moths for part of the night or part of the season, but in some situations they switch to dragonflies just before dawn (R. M. R. Barclay, pers. comm.). The question of specialization by insectivorous bats is further complicated by the problems of measuring the insect prey available to the foraging bats. None of the systems used to date provides a demonstrably accurate sample (Holroyd 1983), and conclusions about prey selection are thus greatly attenuated.

Gleaners, bats taking nonflying prey from surfaces, occur in most parts of the world. Detailed studies of gleaning behavior are available for several species: *Myotis auriculus* (Fenton and Bell 1979), *Cardioderma cor* (Vaughan 1976), *Megaderma lyra* (Fiedler 1979), *Trachops cirrhosus* (Tuttle and Ryan 1981), and *Antrozous pallidus* and *Macrotus californicus* (Bell 1982a; 1982b). Gleaners often take quite large prey; it is impres-

Fig. 11. Ears in the Microchiroptera range from very large to relatively small. Although biologists have tended to identify big-eared bats with a gleaning foraging strategy, *Euderma maculatum* (A), with proportionally the largest ears in the order, pursues flying prey (Leonard and Fenton 1983). The related *Plecotus townsendii* (B) has ears almost as large as those of *E. maculatum*, but at this time details of its foraging behavior are lacking. Reproduced with permission of the University of Toronto Press.

sive to watch a 55 g *Megaderma lyra* catch, subdue, and kill a 30 g laboratory mouse and then pick it up and fly away with it.

The idea that gleaning bats have large ears is common in the literature on bats (e.g., Findley 1976), leading to the generalization that large-eared species are gleaners. Recent work with *Euderma maculatum*, a North American vespertilionid with proportionally the largest ears in the Chiroptera (Fig. 11), clearly shows that this species is not a gleaner and challenges the popular generalization (Leonard and Fenton 1983).

An interesting question about gleaning bats is the extent to which they use echolocation to find their prey. Some species clearly cease to produce orientation calls as they close with prey, while others continue to do so even when relying on cues emanating from the targets (Barclay et al. 1981; Fenton, Gaudet, and Leonard 1983). Responses of prey to the echolocation calls of an attacking bat may account for the bat's using alternate means of orientation, and Werner (1981) demonstrated that sitting moths respond behaviorally to the echolocation calls of hunting bats.

The best-studied bat regularly foraging for fish is the New World noctilionid *Noctilio leporinus* (Suthers 1965; 1967). These bats use echolocation calls to detect fish moving close to the surface, locating ripples associated with the mouths or dorsal fins, and then gaff the fish with their enlarged hind feet. *Noctilio leporinus* does not feed only on fish but also takes other animals, including insects, moving along the surface of the water (Novick and Dale 1971; Howell and Burch 1974). Similar hunting behavior has been reported for some other bats, notably *Noctilio albiventris* (Brown, Brown, and Grinnell 1983) and a variety of *Myotis* spp., which feed along the surface of water and may occasionally include fish in their diet (e.g., Brosset and Deboutteville 1966; Dwyer 1970; Thompson and Fenton 1982). Fish eating is also known from some species of *Megaderma* and from *Nycteris grandis*, but observations of their fishing behavior are lacking.

Larger gleaners often include vertebrates and large invertebrates in their diets. Examples of this pattern of behavior are common in the families Megadermatidae and Phyllostomidae. Vehrencamp, Stiles, and Bradbury (1977) found that the phyllostomid *Vampyrum spectrum* often ate birds, and Fiedler's (1979) laboratory observations on the carnivorous potential of *Megaderma lyra* are supported by field observations showing a variable diet (Advani 1981). Douglas (1967) provided similar observations for the Australian megadermatid *Macroderma gigas*. Tuttle and Ryan (1981) found that some *Trachops cirrhosus* feed heavily on frogs, and this species is also known to take geckoes. There is one very questionable record of a pteropodid, *Hypsignathus monstrosus*, feeding on birds (Van Deusen 1968), but this is not confirmed by more detailed studies of the biology of this species (Bradbury 1977b). I suspect that when more information is available about larger species of gleaners, we will learn that they regularly include smaller vertebrates in their diets. The best example to date is the range of prey taken by *Nycteris grandis* (Fenton, Thomas, and Sasseen 1981) and the occasional exploitation of small bats, *Pipistrellus nanus*, by *Cardioderma cor* (Vaughan 1976).

The three species of vampire bats appear to feed only on blood, and they show a high degree of specialization for this diet (D. C. Turner 1975; Schmidt 1978; Greenhall, Joermann, and Schmidt 1983). Vampires typically approach sleeping victims, make a quick shallow cut with their upper incisors, and drink the blood flowing from the resulting wound. Their saliva contains an anticoagulant, which allows prolonged feeding. Specializations for blood feeding include adaptations for taking off from the ground (Altenbach 1979) and the ability to filter plasma and electrolytes from the blood meal and unload it very quickly (Greenhall, Joermann, and Schmidt 1983). Recent work has also indicated that the common vampire, *Desmodus rotundus*, has heat-sensitive pits around the nose leaf which permit them to locate areas where the blood is close to the surface (Kürten and Schmidt 1982).

There is relatively little information about the usual prey of most vampires, although some are considered to feed more on birds than others. Under artificial conditions, vampires will take blood from a range of animals (Greenhall, Joermann, and Schmidt 1983), but man does not appear to be a preferred food. D. C. Turner (1975) found that *Desmodus rotundus* is as effective in locating cows in estrus as are bulls, and its well-developed Jacobson's organ (Bhatnagar 1980) may be involved in these discriminations. Although vampires are typically labelled as carriers of rabies, evidence for this is wanting (Tuttle and Kern 1981). Vampires, like other mammals, may transmit rabies, but there is no evidence that they are carriers of it.

The three species of vampires are closely related to the phyllostomids and currently are included in that family. The details of their evolution and origin remain unclear, and their history in the light of faunal extinctions in South America (e.g., Simpson 1980) is obscure. Fossil *Desmodus* are known from Florida (Greenhall, Joermann, and Schmidt 1983) and *D. rotundus* seems to have undergone a dramatic increase in population following the introduction of many domestic mammals to the New World.

Plant Products

The plant product most commonly exploited as food by bats is fruit. In the Neotropics fruit-eating bats are usually phyllostomids, although *Antrozous pallidus*, a vespertilionid, will also take fruit (D. J. Howell 1980). In the Old World tropics, fruit-eating bats are the Pteropodidae, providing an interesting example of convergent evolution between the two groups. Fruits may be consumed on the plants where they grow, or picked and taken to night roosts for consumption. In the Neotropics Fleming (1982) has identified two different strategies of fruit eating.

Some species search widely for fruit, quickly locating new supplies, while others visit predictable patches. Comparable details for the Old World situation are lacking. There is evidence of specialization and resource partitioning in some communities of fruit bats, in both the Old World and the New World (D. W. Thomas 1982; Fleming 1982, respectively). In the Old World the lack of ability to echolocate by most fruit eaters seems to preclude the regular use of insects as a source of protein. In the Neotropics, the frugivorous phyllostomids are commonly assumed to supplement their diets with insects as a source of protein. D. W. Thomas (1982) has suggested that low protein levels in fruits have contributed to co-evolution between bats and plants in the Old World tropics.

Nectar and pollen are the other plant products commonly consumed by bats. In the tropics there are long-snouted, long-tongued species adapted for visiting flowers and extracting nectar and pollen (e.g., Gould 1978a; D. J. Howell 1979). A large number of plants are chiropterophilus, with flowers adapted to attract and exploit bats (Heithaus 1982). D. J. Howell (1979) has shown that at least the phyllostomid *Leptonycteris sanborni* obtains protein by digesting pollen, partly because it drinks its own urine to create an acidic environment in its digestive tract. An important aspect of this situation is allogrooming by *L. sanborni* which roost together (D. J. Howell 1979). There is a spectrum of dependence upon flowers and their products among New World flower bats, involving morphology, behavior, and echolocation (D. J. Howell 1974).

In the Old World tropics, Gould (1978a) has shown a close association between the pteropodid *Eonycteris spelaea* and some plants, but there is, at this time, no clear indication of how the specialized pteropodids obtain their protein. D. W. Thomas (1982) failed to find evidence that pollen constitutes a regular and significant role in the diets of the pteropodids he studied in Ivory Coast, but some of these bats visit flowers in other parts of Africa (e.g., Ayensu 1974).

D. J. Howell's (1980) report of the animal-eating *Antrozous pallidus* taking fruit, combined with Daniel's (1979) observations on the endemic New Zealand *Mystacina tuberculata* and Gardner's (1977) review of the food habits of phyllostomids, clearly indicates that flexibility in diet is a feature of many bats. Gillette (1975) stressed the importance of duality in diets as a feature of many bats and noted its importance in their evolution. When this is combined with opportunistic feeding behavior

(Fenton and Morris 1976; Gould 1978b; Vaughan 1980), the diversity of the Chiroptera becomes clear. The ability of bats to alter their feeding behavior in accordance with available resources and to switch from animal to plant products is an important trait. These features of the behavior of bats have important implications for, and should be considered in the light of, studies of their communication.

Roosts

The roosting habits of bats were reviewed recently by Kunz (1982a). Since roosts play different roles in the lives of bats, many species use a variety of them. Day roosts serve as diurnal retreats, night roosts as resting, eating, or digestion centers, and hibernacula as places to pass extended inclement periods. The adaptations for flight in bats result in a relatively thin profile through the chest, permitting use of narrow spaces and suggesting that secure roosts have been an important factor in their evolution.

Bats may be divided into several categories based on the kinds of roosts they usually occupy. Before considering this in more detail, it is important to realize that for many species roosts remain unknown. Furthermore, there is little information about the range of roosts occupied by a species, or the roost repertoire of individuals. Some species roost predictably and repeatedly in the same situations, and some individuals always return to the same roosts. In other species, however, individuals move unpredictably between different roosts, sometimes occupying the same kind of site, but rarely the same site two days in a row (Table 2).

The phyllostomid *Carollia perspicillata*, studied in the field by radio-tracking (Heithaus and Fleming 1978), tended repeatedly to use the same roost, while another phyllostomid, *Phyllostomus hastatus*, consistently used the same roost over a two- or three-year period (McCracken and Bradbury 1981). In contrast, other species, such as the phyllostomid *Vampyrodes caraccioli*, remained in the same general area but rarely occupied the same roost two nights in a row (Morrison 1980). Shifting from roost to roost is a common phenomenon in some species; others are more philopatric (Table 2). In some situations, changing roosts reflects roost availability. For example, the thyropterid *Thyroptera tricolor* is limited in the time it can roost in unfurled leaves by the leaf's opening (Findley and Wilson 1974). The significance of roost-switching behavior and its incidence among bats remain to be determined (Fenton 1983b).

Some bats roost in hollows, others in crevices or foliage. Hollows may be in rock (caves, mines), trees, or buildings. Species occupying crevices may select sites in rock, around trees (for example, under loose bark), or even under rocks on the ground. Many species commonly exploit crevices associated with buildings or other structures such as culverts or bridges. Species roosting in foliage may be conspicuous, as is the case in the camps of many species of flying foxes (*Pteropus* spp.), or cryptic. A few bats exploit more bizaare roosting situations; e.g., two species of vespertilionids (*Tylonycteris* spp.) in Malaysia roost in the internodes of bamboo stems, entering the hollows via holes bored by beetles. Several species of bats have adhesive discs on their wrists and ankles, permitting them to gain purchase on slippery surfaces, and this in turn correlates with roosting in unfurled leaves. Species which roost under rocks often have protective warts on their forearms (Peterson 1965b). Although bats do not build nests, some species modify leaves to make tents, a pattern of behavior known from some phyllostomids and one pteropodid (Kunz 1982b).

Shelter from sun, rain, and predators is a feature common to many of the diurnal roosts occupied by bats, although many foliage-roosting species, particularly flying foxes, gain little or no protection from the elements in their roosts. Ambient light conditions within roosts range from total darkness to full sunlight, and temperatures range from over $35°C$ to less than $5°C$ in some hibernacula. High concentrations of bats, with associated accumulations of urine and feces, make some roosts oppressive. For example, K. M. Howell (1980) found that within a kaolin mine occupied by thousands of the hipposiderid *Triaenops persicus* there was not enough oxygen to sustain the flame of a pressure lamp. Bats show interesting adaptations permitting them to tolerate high concentrations of ammonia (Studier 1969).

Superimposed on the variety of roosts occupied by bats is a considerable range of roosting habits. Some roosts are occupied only by conspecifics, others by mixtures of species. In some cases roosts may house only one sex-age group, for example, a cluster of bachelor males, although more commonly in cave-roosting species bachelor groups are confined to different parts of the cave (McCracken and Bradbury 1981; McWilliam 1982). Some species roost alone or only with their current young; others aggregate, reflecting the limited (limiting?) nature of

TABLE 2. Patterns of use of day roosts by different bats

SPECIES	ROOST SITE	STUDY TECHNIQUE	ROOST FIDELITY; PATTERNS OF MOVEMENT BETWEEN ROOSTS	SOURCE
Pteropodidae				
Epomophorus gambianus (in Zimbabwe)	F[1]	rt[2]	frequent moves between adjacent or close trees	D. W. Thomas & Fenton 1978
Epomophorus gambianus (in Côte d'Ivoire)	F	rt	use of a variety of sites within a small area (300 m diameter)	D. W. Thomas 1982
Hypsignathus monstrosus	F	rt	shift between sites at short intervals	Bradbury 1977a
Epomops buettikoferi	F	rt	use same roost unless disturbed	D. W. Thomas 1982
Emballonuridae				
Rhynchonycteris naso	T[3]	rb[4]	3 to 6 roost sites, move between them at intervals	Bradbury & Vehrencamp 1976
Saccopteryx bilineata	T	rb	use one roost for a long period of time	Bradbury & Vehrencamp 1976
Saccopteryx leptura	T	rb	set of roosts, move between them at intervals	Bradbury & Vehrencamp 1976
Phyllostomidae				
Carollia perspicillata	C/F[5]	rt	most bats consistently used one roost	Heithaus & Fleming 1978
Phyllostomus hastatus	C	rt	consistently used the same roost over 2 or 3 years	McCracken & Bradbury 1981
Vampyrum spectrum	H[6]	rt	used one roost	Vehrencamp, Stiles, & Bradbury 1977

Species				
Vampyrodes caraccioli	F	rt	remained in same general area but used a range of roost sites, rarely same roost two nights consecutively	Morrison 1980
Artibeus lituratus	F	rt	as for *V. caraccioli*	Morrison 1980
Artibeus jamaicensis	H/F	rt	fidelity depended upon nature of roost, some movement between roosts	Morrison 1979
Vespertilionidae				
Eptesicus fuscus	B[7]	rt	frequent moves between roosts over large area (over 1 km diameter)	Geggie 1983
Scotophilus leucogaster	H	rt	frequent moves between roosts from day to day over small area	Fenton 1983b
Antrozous pallidus	C	o[8]	frequent shifting from roost to roost	Vaughan & O'Shea 1976

[1]foliage; [2]radio-tracking; [3]tree trunk; [4]reflective band; [5]C-cave; [6]hollow tree; [7]building; [8]observation

some roost resources; other species roost in modest or even large numbers in strict social organization. Some species use their collective body heat to raise the temperatures in otherwise unsuitable sites, making them exploitable as nursery roosts (e.g., *Myotis grisescens;* Tuttle 1975).

REPRODUCTION

In ecological terms based on fecundity, bats are K strategists, producing relatively few young per litter and few litters per year. Reproduction in bats is effectively reviewed by Racey (1982) and Gustafson and Weir (1979). Most bats have one or two young per litter, and the largest litters, up to four young, are known from the vespertilionid *Lasiurus borealis* (Barbour and Davis 1969). Temperate bats are typically monestrus, while some tropical species are polyestrus, usually with two litters per year (e.g., Fleming 1971; 1973), or sometimes more (e.g., Beck and Lim 1973). Some species, such as the vespertilionid *Myotis nigricans,* may vary between two and three litters annually, depending upon prevailing conditions (Wilson and Findley 1970), and in many species polyestry is facilitated by a postpartum estrus (Myers 1977).

We know most about the reproductive biology of temperate bats, reflecting the distribution of biologists with the resources to study reproduction in detail. The temperate species show patterns of reproduction strongly influenced by the seasonal environments in which they occur. Typically the temperate vespertilionids and rhinolophids mate in the autumn or winter and postpone fertilization until females depart from hibernation in the spring. This pattern of delayed fertilization is also known from some tropical vespertilionids, upsetting the obvious association between hibernation and delayed fertilization. Sperm storage by females has been relatively little studied (Racey 1982), and for most of the species for which there is clear evidence of this phenomenon, there are no data on the mating systems involved (Fenton, in press).

Other bats show patterns of reproduction which permit mating at one time of the year and postponement of birth until more suitable seasons. In some phyllostomids and vespertilionids this delay is achieved by delayed implantation or delayed development, both of which are known from bats from a range of families (Table 3).

Strategies which permit mating at one time of the year and birth at some subsequent time beyond the range of "normal" gestation periods for bats are important to biologists studying chiropteran reproductive behavior. The situation is further complicated by bats which show periods of gestation extendable by their entry into torpor (Racey 1982).

The period of lactation is variable among bats, with young *Myotis lucifugus* starting to take solid food at about 18 days of age, and young *Desmodus rotundus* being weaned when they are more than 9 months old (Schmidt and Manske 1973). In some cases, for example in *D. rotundus* or *Antrozous pallidus* (Brown 1976), some young bats are fed a combination of milk and solid foods, but in most instances it is not clear whether solid foods appearing in the diets of young were obtained from the parent or independently. It is well known that in captivity *Desmodus rotundus* young are fed regurgitated food by adults (Greenhall, Joermann, and Schmidt 1983), and Racey (1982) reported two instances of female bats bringing insects to their young, one involving the molossid *Molossus ater*, the other the emballonurid *Coleura afra*. Brown, Brown, and Grinnell (1983) reported that captive female *Noctilio albiventris* used their cheek pouches to bring insects to their unweaned young. In spite of extensive observations within a maternity colony of *Myotis lucifugus* under natural conditions, Thomson (1980) was unable to establish whether or not this species brought insect food to their young during the process of weaning. In *M. lucifugus* appearance of insect remains in the digestive tract coincides with the eruption of the permanent dentition and the ability to fly (Fenton 1970).

There is clearly variation in mother-young relationships among different species of bats, and the most striking feature of these interactions is our current lack of information about them. As we shall see, interactions between mothers and their young involve some of the most complex vocalizations in the repertoires of bats.

For species of bats which form nursery colonies there are several levels of interactions between participating females. In some species, for example *Myotis lucifugus*, females aggregate in roosts which serve as nurseries. In these situations, these bats, and other species such as *Myotis yumanensis*, adjust their positions within the roost and their proximity to other bats according to prevailing temperature conditions, clustering when it is cold and spreading out more when it is hot (Licht and Leitner 1967). Species such as *Myotis grisescens* aggregate in large numbers and rely on their collective body heat to raise the ambient temperatures in otherwise too-cool roosts, turning them into

TABLE 3. Delayed fertilization, implantation, and development in the Chiroptera (data from Racey 1982, and Fenton, in press b)

FAMILY SPECIES	DELAY STRATEGY	DURATION OF DELAY	SOURCE
Pteropodidae			
Eidolon helvum	delayed implantation	3 months	Mutere 1967; Fayenuwo & Halstead 1974
Cynopterus sphinx	delayed development[1]	30 days	Krishna 1978
Emballonuridae			
Taphozous longimanus	delayed development[1]	20 days	Krishna 1978
Phyllostomidae			
Macrotus californicus	delayed development	4.5 months	Bradshaw 1961, 1962; Bleier 1975
Artibeus jamaicensis	delayed development	3 months	Fleming 1971
Rhinolophidae			
Rhinolophus rouxi	variable rate of post-implantation growth		Ramakrishna & Rao 1977
Rhinolophus ferrumequinum	delayed fertilization	over winter	Racey 1975
Rhinolophus hipposideros	delayed fertilization	over winter	Racey 1975
Vespertilionidae			
Myotis albescens	delayed development[1]	unknown	Myers 1977
Myotis lucifugus	delayed fertilization	138 days	Wimsatt 1944
Myotis sodalis	delayed fertilization	68 days	Gates 1936
Myotis tricolor	delayed fertilization	unknown	Bernard 1980
Myotis nattereri	delayed fertilization	unknown	Racey 1975
Myotis daubentoni	delayed fertilization	unknown	Racey 1975

Pipistrellus pipistrellus	delayed fertilization	151 days	Racey 1973
Pipistrellus abramus	delayed fertilization	175 days	Haraiwa & Uchida 1956
Pipistrellus ceyloncus	delayed fertilization	16 days	Gopalakrishna & Madhavan 1971; Racey 1979
Nyctalus noctula	delayed fertilization	198 days	Racey 1973, 1975
Eptesicus fuscus	delayed fertilization	156 days	Wimsatt 1944
Eptesicus pumilus	delayed fertilization	unknown	Green 1965
Eptesicus regulus	delayed fertilization	unknown	Kitchener & Halse 1978
Eptesicus furinalis	delayed development[1]	unknown	Myers 1977
Tylonycteris pachypus	delayed fertilization	21 days	Racey 1979
Tylonycteris robustula	delayed fertilization	unknown	Racey 1979
Scotophilus heathi	delayed fertilization	unknown	Krishna & Dominic 1978; Gopala-krishna & Madhavan 1978
Chalinolobus gouldii	delayed fertilization	unknown	Kitchener 1975
Lasiurus ega	delayed fertilization	unknown	Myers 1977
Plecotus townsendii	delayed fertilization	unknown	Pearson, Koford, & Pearson 1952
Antrozous pallidus	delayed fertilization	unknown	Orr 1954
Miniopterus schreibersi	delayed implantation	2.5–5 months	van der Merwe 1979; Richardson 1977; Wallace 1977; Peyre & Herlant 1967; Courrier 1927
Miniopterus fraterculus	delayed implantation	2.5 months	Bernard 1980
Miniopterus australis	delayed development	4–5 months gestation	Medway 1971; Richardson 1977
Natalidae			
Natalus stramineus	delayed development	unknown	Wimsatt 1975

[1] suggested but not clearly demonstrated.

TABLE 4. Growth rates and age at sexual maturity in bats (from Tuttle and Stevenson 1982)

FAMILY SPECIES	GROWTH RATE FOREARM MM/DAY	GROWTH RATE WEIGHT G/DAY	AGE AT SEXUAL MATURITY ♂♂	AGE AT SEXUAL MATURITY ♀♀	SOURCE
Pteropodidae					
Hypsignathus monstrosus	0.7	1.3	18 mo	6 mo	Bradbury 1977b
Rousettus spp.	1.5	—	—	—	Kulzer 1969
Rousettus leschenaulti	—	—	15 mo	5 mo	Gopalakrishna & Choudhari 1977
Dobsonia moluccensis	—	—	>2 years	—	Dwyer 1975
Pteropus ornatus	—	—	~16 mo	~16 mo	Asdell 1964
P. poliocephalus	—	—	~16 mo	~16 mo	Nelson 1965
P. scapulatus	—	—	~16 mo	~16 mo	Nelson 1965
Rhinopomatidae					
Rhinopoma kinneari	—	—	~16 mo	~16 mo	Anand 1965
Emballonuridae					
Balantiopteryx plicata	—	—	—	1 year	Bradbury & Vehrencamp 1977
Rhynchonycteris naso	—	—	—	>1 year	Bradbury & Vehrencamp 1977
Saccopteryx bilineata	—	—	—	1 year	Bradbury & Vehrencamp 1977
Saccopteryx leptura	—	—	—	1 year	Bradbury & Vehrencamp 1977

Taphozous melanopogon	—	—	6 mo	—	Brosset 1962
Megadermatidae					
Cardioderma cor	—	—	16 mo	16 mo	Matthews 1941
Megaderma lyra	—	—	15 mo	19 mo	Ramakrishna 1951; Brosset 1962
Hipposideridae					
Hipposideros bicolor	—	—	~1 year	~1 year	Brosset 1962
H. fulvus	—	—	18–19 mo	18–19 mo	Madhavan & Gopalakrishna 1978
H. speoris	—	—	~16 mo	~16 mo	Brosset 1962
Rhinolophidae					
Rhinolophus hipposideros	1.0	—	15 mo	3–15 mo	Rybaur 1971; Sluiter 1960; Gaisler & Titlbach 1964; Gaisler 1965
R. euryale	—	—	2.25 years	1.25–2.25 years	Dinale 1968
R. ferrumequinum	—	—	1.25–4.25	1.25–3.25 years	Asell 1964; Dinale 1964
R. rouxi	—	—	—	>1 year	Brosset 1962
Phyllostomidae					
Carollia perspicillata	0.8	0.3	~16 mo		Kleiman & Davis 1979
Macrotus californicus	—	—	—	3–4 mo	Bradshaw 1961
Artibeus jamaicensis	—	—	—	8–11 mo	Fleming, Hooper, & Wilson 1972
Phyllostomus hastatus	—	—	—	~16 mo	McCracken & Bradbury 1981

TABLE 4. (*Continued*)

FAMILY SPECIES	GROWTH RATE		AGE AT SEXUAL MATURITY		SOURCE
	FOREARM MM/DAY	WEIGHT G/DAY	♂♂	♀♀	
Vespertilionidae					
Myotis adversus	0.6	—	—	—	Dwyer 1970
M. austroriparius	—	—	~10–15 mo	~3–10 mo	Rice 1957
M. emarginatus	—	—	—	3 mo	Sluiter & Bouman 1951
M. grisescens	—	0.2–0.4	16 mo	16 mo	Tuttle 1975; Guthrie 1933; R. E. Miller 1939
M. lucifugus	1.6	0.34	~16 mo	~3–16 mo	Kunz & Anthony 1982; Herd & Fenton 1983
M. macrodactylus	0.9	—	—	—	Maeda 1976
M. myotis	2.0	—	15 mo	3 mo	Kratky 1970; Sluiter & Bouman 1951; Sluiter 1961
M. mystacinus	—	—	—	3–15 mo	Sluiter 1954
M. nigricans	—	—	2.5–4 mo	3–4 mo	Wilson 1971; Myers 1977
M. thysanodes	1.5	0.3	—	—	O'Farrell & Studier 1973
M. velifer	1.6	0.4	~16 mo	~4 mo	Kunz 1973; Hayward 1970
Eptesicus fuscus	0.8–1.4	0.3–0.47	4 mo	4–16 mo	W. H. Davis, Barbour, & Hassell 1968; Kunz 1974; Burnett & Kunz 1982; Christian 1956

Species					Reference
E. serotinus	1.2	0.9	—	—	Kleiman 1969
E. furinalis	—	—	<1 year	<1 year	Myers 1977
Lasiurus cinereus	0.4	—	—	—	Bogan 1972
L. ega	—	—	<1 year	<1 year	Myers 1977
Nyctalus lasiopterus	0.9–1.6	0.6–1.0	~16 mo	~4 mo	Maeda 1972; Maeda 1974
N. noctula	0.9	0.6	3–15 mo	3 mo	Kleiman 1969; Kleiman & Racey 1969; Gaisler, Hanak, & Dungel 1979
Pipistrellus pipistrellus	0.7–0.8	0.1–0.42	16 mo	4 mo	Kleiman 1969; Rakhmatulina 1972; Asdell 1964; Racey 1974
Scotophilus temmincki	—	—	~9 mo	~9 mo	Gopalakrishna 1947
Tylonycteris pachypus	—	—	3 mo	6 mo	Medway 1972
T. robustula	—	—	3 mo	6 mo	Medway 1972
Plecotus auritus	—	—	—	~16 mo	Stebbings 1966
P. rafinesquii	—	—	~16 mo	—	Jones & Suttkus 1975
P. townsendii	1.2	—	~4–18 mo	~4 mo	Pearson, Koford, & Pearson 1952
Antrozous pallidus	1.5	0.3–0.4	—	—	R. B. Davis 1969
Miniopterus australis	—	—	—	~16 mo	Dwyer 1968
M. schreibersi	0.7	0.3	~16 mo	~16 mo	Dwyer 1963; Brosset 1962
Molossidae					
Tadarida brasiliensis	0.7–1.0	0.4	~16–22 mo	9 mo	Pagels & Jones 1974; Short 1961; Sherman 1937

suitable nurseries. As Tuttle and Stevenson (1982) pointed out, species exposed to the harshest conditions may be forced to cooperate to a greater extent and should show the highest synchrony of parturition.

There is also variation in the degree of development of bats at birth. Among the pteropodids, the young are smaller at birth relative to the females, but more advanced developmentally (Orr 1970). Within the Microchiroptera, phyllostomids tend to be most advanced at birth (Kleiman and Davis 1979), and W. B. Davis (1944) reported that in the emballonurids young tended to be largest at birth relative to their mothers. In general, baby bats tend to be huge at birth compared to their mothers. Breech births are common, perhaps to minimize the chances of the young becoming tangled in the birth canal. The thumbs and hind feet of baby bats are almost adult in size, permitting the young to cling to their mothers.

There is variability in the rates of growth of young bats (Table 4), but in all cases newborn bats are highly dependent upon their mothers at least for the period until they can fly. The details of weaning and postweaning interactions between females and their young also have been little studied. There is also considerable variation in the age at which different bats reach sexual maturity (Table 4).

Bats exhibit a range of mating systems, from species presumed to be monogamous to others which are polygynous, and at least one species forms leks (Bradbury 1977a; 1977b). Interactions between male and female bats often involve complex communication signals, which may be visual, vocal, or olfactory.

Energy and Activity

McNab (1982) pointed out that bats are commonly considered to have poor temperature regulation, to feed on insects, and to enter hibernation, even though these are attributes of temperate species. Some bats have the ability to function over a wide range of temperatures. For example, torpid *Myotis lucifugus* have heartbeat rates of about 8 per minute, depending upon the ambient temperature; when the bat is active at 35°C the rate is over 200 per minute, and when the bat is flying it is over 1300 per minute (Kallen 1977). Only bats in the families Vespertilionidae and Rhinolophidae are known to hibernate or enter torpor; most families and most species of bats do not share this attribute. The dietary diversity in the Chiroptera has already been reviewed.

Most tropical species cannot tolerate extended low temperatures or low body temperatures, and the environmental thermal regime probably is a prime factor in determining the range of many bats (McNab 1982). Species in the families Nycteridae and Hipposideridae, for example, show continuous high levels of activity and metabolism which must be sustained by regular supplies of food. High tuning of their metabolic rates is partly responsible for their vulnerability to death in untended traps, particularly during cool dry conditions. McNab (1973) has previously reviewed the importance of thermal regime on the distribution of vampire bats, based on our knowledge of their thermoregulatory and other abilities.

Bats appear to face a perpetual series of problems associated with heat, either too much or too little of it. Flying bats apparently need to disperse heat built up by contractions of their flight muscles, while resting bats with high metabolic rates and large surface areas must conserve heat. Direct anastomoses between small arteries and small veins in the wings (Kallen 1977) probably facilitate unloading of heat by flying bats, and folding of wings upon landing effectively reduces the surface area from which bats lose heat. In many species large ears can be folded back, further reducing surface area and cutting down on heat loss.

An important energy-saving strategy employed by vespertilionids and rhinolophids in temperate and tropical areas is an adjustable thermostat, manifested as the ability to enter torpor. At least in the vespertilionid *Myotis lucifugus*, torpor of bats in summer is distinct from the torpor of hibernation. During summer bats cannot spontaneously arouse from torpor; they can from winter torpor (Menaker 1962). Careful selection of roosts according to the prevailing microclimate permits vespertilionids and rhinolophids to accommodate their metabolic rates (and energy consumption) to prevailing conditions. There are, as we have seen, a variety of roosting strategies, from species which adjust their degree of physical contact (clustering) according to the ambient temperature (Licht and Leitner 1967), to those which use their collective body heat to make effective nurseries (Tuttle 1975).

Torpor, an indication of the variable thermostat rather than a reflection of poor temperature regulation, is a means by which small animals with large surface areas and high metabolic rates can conserve energy and prosper in marginal habitats which periodically support high levels of prey populations (cf. Heinrich 1979). It is inappropriate to consider bats as poor thermal regulators.

The activity patterns of bats are strongly influenced by ambient temperatures (McNab 1982; Erkert 1982) and other circadian factors (Erkert 1982). Evening departures from roosts are usually regulated by ambient levels of light, and many species depart earlier on overcast than on clear nights (Erkert 1982). Since insect activity is greatly reduced below ambient temperatures of 10°C (C. B. Williams 1940), flight activity of insectivorous bats is also reduced under these conditions. At higher temperatures bat activity is influenced by weather conditions and moonlight. Heavy rain depresses the flight activity of some bats, but others continue to forage in the rain, exploiting insects that remain active (Erkert 1982). Some authors have related lower flight activity by echolocating bats in the rain to interference with echolocation partly due to atmospheric attenuation of high-frequency sounds, but a more plausible explanation involves temperature regulation. Wet wings and fur would affect rates of heat loss and could upset the delicate temperature balance of a flying bat (Fenton, Boyle, et al. 1977). Neither rain nor wind completely inhibits the foraging activity of bats.

Moonlight strongly inhibits and/or alters the flight and foraging activity of many species of bats (e.g., Fenton, Boyle, et al. 1977; Morrison 1978; Erkert 1982), but some others are not demonstrably affected (Leonard and Fenton 1983). There is no obvious relationship between moonlight and bat flight activity in terms of energy, but predators, particularly birds such as bat hawks and owls, are usually invoked as important influences. Clear evidence that the threat of predation is the most plausible explanation for changes in behavior in bright moonlight is wanting.

Energetics strongly influence the lives of bats, from the food they consume to the locations where they roost. Ambient temperatures clearly influence the nature of vocalizations of bats which enter torpor, and in temperate regions species mating in the fall (apparently the majority) face the conflicting demands of preparation for winter and mating at a time when food supplies are declining. The methods used by bats to resolve their energy demands have important implications for social structure of roosting and feeding populations, and for associated patterns of communication.

II

Media for Communication

VISION

The eyes of bats are typically mammalian in structure and operation, but there are important differences between the two suborders. The retinas of either group are dominated by rods, although cone-like cells appear in some species (Suthers 1970a). There is no evidence of a fovea in the eyes of bats examined to date, and a tapetum lucidum is only known from the eyes of Megachiroptera. In general, the eyes of bats are adapted for operation under conditions of low light, although variation in this trait is to be expected reflecting differences in roost-site selection. Information from the retina is relayed by the optic nerve to the lateral geniculate body and thence to the visual cortex, again in typically mammalian pattern.

The unique feature of the megachiropteran eye is the small conical projections, choroidal papillae or cones, whose contours are followed by the retina. There is also a unique interdigitation of the retina and the choroid (Kolmer 1924). Within the Megachiroptera there is variation in the degree of development of the choroidal papillae. The papillae all point to the nodal point of the dioptric apparatus, preventing shadowing of one papilla by another; this arrangement gives a toothlike appearance in a section through the eye. Although these papillae have been proposed as a mechanism to increase the number of receptor cells (e.g., Pedler and Tilley 1969), because the long axis of the receptor cells on the papilla is parallel to the central axis of the papilla changes in the choroidal surface area per se do not alter the number of receptors. What is important in this context is the projected surface area, not the actual surface area (Suthers 1970b). The significance of the choroidal papillae remains unclear.

Although earlier investigators had suggested that the ciliary muscles in the Megachiroptera were poorly developed and that these species would thus lack good accommodation ability (Suthers 1970a), recent work has challenged this view. Although the ciliary muscles are not as developed as they are in primates in general, they are as well developed in Megachiroptera as they are in some nonprimates, and these bats have a clearly active accommodation mechanism (Murphy et al. 1983).

Some early work had also suggested that bats in general were far-sighted (hypermetropic). Examination of relatively small eyes by means of retinoscopy, however, often suggests hypermetropic conditions because of reflection from the inner surface of the retina; in other words, in many cases this impression is the result of an error in measurement (Glickstein and Millodot 1970).

In both the megachiropterans and the microchiropterans there is considerable variation in the actual size of the eyeball, and in both groups the cornea tends to be large and has a greater radius of curvature than the eyeball (Suthers 1970a). This condition is typical of the eyes of nocturnal mammals. In both groups of bats, the lens tends to be thick, not spherical, and about 50 percent of the eye's axial diameter. Some of the microchiropterans are unable to close the pupil in bright light, which would have important repercussions for their roosting habits.

Acuity, as indicated by minimum separable angles, varies among different species of bats (Table 5). Of particular interest in this context is variation in acuity among different species of insectivorous bats, *Macrotus californicus* and *Antrozous pallidus* in Table 5. In both suborders visual acuity deteriorates with decreasing light levels more slowly than man's so that in dim light bats can usually see better than humans (Suthers 1970a). Their thresholds are generally similar to those reported for owls (Neuweiler 1967). In some Microchiroptera there is no clear threshold at which bats switch from orientation by echolocation to orientation by vision (Fig. 12), but there is a hyperbolic relationship (Bell 1982b). Visual acuity has been measured by tasks which may or may not involve flying animals (e.g., Suthers 1970a; Bradbury and

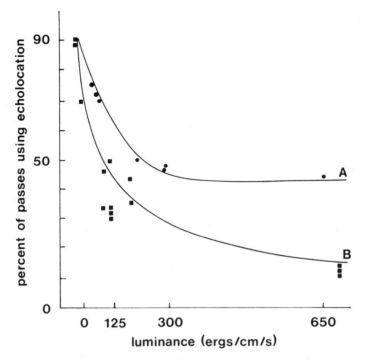

Fig. 12. A comparison of the incidence of echolocation calls under increasing light intensities for foraging *Macrotus californicus* (A) and *Megaderma lyra* (B) shows how both of these species tend to cease production of echolocation calls when there is adequate light for orientation by vision. In the experiments (data from Bell 1982b and M. S. Kelly, pers. comm.) no sound cues emanating from the targets were available for the bats.

Nottebohm 1969; Masterson and Ellins 1974). It is clear that many species of bats distinguish visually between different patterns (Suthers 1970a).

Among Microchiroptera there is divergence from typical rhodopsin absorption spectra and a second photoreceptor class and pigment (Hope and Bhatnagar 1979a). Along the gradient from *Eptesicus fuscus* through *Desmodus rotundus* and *Artibeus jamaicensis* to *Carollia perspicillata* there are progressively more photoreceptors (Hope and Bhatnagar 1979a) and electroretinography suggests in these species the retinas are progressively more light-tolerant, indicating increasing capacity for functioning in brighter light (Hope and Bhatnagar 1979b). In terms of

TABLE 5. Visual acuity in different bats

FAMILY SPECIES	MINIMUM VISIBLE ANGLE	SOURCE
Emballonuridae		
Saccopteryx leptura	42'	Chase 1972
Rhinolophidae		
Rhinolophus hildebrandti	>1°40'	Bell, pers. comm.
Hipposideridae		
Hipposideros ruber	1°40'	Bell, pers. comm.
Cloeotis percivali	>1°40'	Bell, pers. comm.
Nycteridae		
Nycteris thebaica	1°40'	Bell, pers. comm.
Phyllostomidae		
Macrotus californicus	8'16"	Bell 1982b
Carollia perspicillata	16'	Chase 1972
Anoura geoffroyi	42'	Chase 1972
Desmodus rotundus	42'	Chase 1972
Vespertilionidae		
Myotis lucifugus	3°	Chase 1972
Eptesicus fuscus	2°	Bell 1982b
Eptesicus capensis	50'	Bell, pers. comm.
Pipistrellus nanus	1°40'	Bell, pers. comm.
Antrozous pallidus	30'	Bell 1982b
Molossidae		
Tadarida ansorgei	>1°40'	Bell, pers. comm.
Molossus ater	10°	Chase 1972.

tolerance of light, *A. jamaicensis* is more advanced than *C. perspicillata* and often roosts in brighter situations.

In summary, bats in either suborder have eyes adapted for operation under conditions of low light, although there is variation in tolerance of bright conditions. Visual acuity varies between different species of bats, which is not surprising given their diversity and the many conditions under which they operate. There is no evidence that any species of bats should be considered blind. The optical properties of the eyes of echolocating bats are consistent with the hypothesis that vision is used for detecting objects beyond the relatively short range of echolocation (Suthers and Wallis 1970).

Receiver—Olfactory Epithelium

The olfactory system of bats is typically mammalian in arrangement. Olfactory epithelium housing receptor cells is located on some turbinals or scroll bones (can include nasoturbinals, maxilloturbinals, and ethmoturbinals). Some turbinals arise from the cribriform plate (of the ethmoid), others from the lateral nasal wall. Ectoturbinals, which can be naso-, maxillo, -or ethmoturbinals, are more dorsolateral, while endoturbinals tend to be more ventromedial. The cribriform plate of the ethmoid bone is perforated by nerves coming from the receptors. Information is carried to the olfactory bulb, which houses glomeruli, and from there impulses travel via mitral cells to the olfactory cortical system and the prepyramidal cortex, or via tufted cells to the amygdala and related regions of the forebrain.

The cribriform plate in the Chiroptera shows varying degrees of perforation, reflecting differences in plate area, amounts of sensory epithelium, and olfactory acuity. The cribriform plates of the Megachiroptera tend to be more perforated than those of the Microchiroptera, and within the Microchiroptera, frugivorous species have more perforations of the plate than insectivorous ones. Intermediate levels of perforation are common, particularly among the more omnivorous of the phyllostomids (Suthers 1970a). In general, increased perforations correlate with increased epithelial area and increased size of olfactory bulb (Bhatnagar and Kallen 1974a).

The olfactory epithelium contains two types of bipolar receptor cells, and there is variability within and between species in the amount of epithelium (Suthers 1970a). Some of this variability is related to the arrangement of the ecto- and endoturbinals (Kamper and Schmidt 1977). Again, frugivorous species tend to have more epithelium than insectivorous ones. For example, Bhatnagar and Kallen (1975) compared the nasal epithelium of the frugivorous *Artibeus jamaicensis* with the insectivorous *Myotis lucifugus*. In the former, 55.9 percent of the nasal cavity was coated with sensory epithelium, compared to 28.9 percent in *M. lucifugus*, meaning that *A. jamaicensis* has 16 times the surface area of sensory epithelium.

The general organization of the olfactory bulb of bats is similar to that of *Mus musculus,* but in bats the bulbs are relatively and absolutely smaller (Suthers 1970a). In general, the larger (⩽ 2 mm) bulb diameter means a better sense of smell, and is usually typical of frugivorous species. The ratio of bulb diameter to cerebral hemisphere size, however, does not provide an accurate prediction of olfactory ability. There is considerable variation in the size of olfactory bulbs within the Chiroptera, but in general, insectivorous species have less conspicuous bulbs than frugivorous ones (Schneider 1957).

Overall, a reflection of olfactory ability may be control over the airstream, which is better in *Artibeus jamaicensis* than it is in *Myotis lucifugus;* also there is in *A. jamaicensis* a clear demarcation of olfactory and respiratory portions of the olfactory cavity (Bhatnagar and Kallen 1974b). Despite the greater area of sensory epithelium in *A. jamaicensis,* however, the number of mitral cells in the olfactory bulbs and olfactory nerves of both bats is similar.

Data on olfaction demonstrate clear differences in anatomy, which presumably correlate with differences in sensitivity. Nevertheless, relatively few species have been studied in detail, and most of the analysis has focused on dietary rather than communicatory differences.

Receiver—Vomeronasal Organ

Bhatnagar (1980) has recently reviewed the incidence of the vomeronasal organ complex in adult bats (see also Suthers 1970a; Cooper and Bhatnagar 1976). Within the Chiroptera there exists a complete range of the vomeronasal organ complex, from well developed to totally absent. This structure, also known as Jacobson's organ, may be a prominent feature in the anteroventral nasal septum when it is present. It is a bilateral structure containing epithelium generally similar to that of the olfactory system and connects, via the vomeronasal nerve, to the accessory olfactory bulb.

Although there is a spectrum of development of this feature, there is considerable homogeneity in some instances. For example, the vomeronasal organ has not been found in the Megachiroptera examined to date, and within some pairs of species, such as the vampires *Desmodus rotundus* and *Desmodus* (= *Diaemus*) *youngi,* the structure may be very similar (Bhatnagar 1980). Six conditions have been identified: (i) a patent duct and vomeronasal organ opening into it; (ii) a patent duct only with no vomeronasal organ; (iii) a vomeronasal organ only; (iv) blind ducts; (v) blind ducts on one side; (vi) no sign of duct or organ.

There is no clear separation of groups of bats by diet on the basis of vomeronasal organ structure, although the organs are best developed in some of the vampire bats (Bhatnagar 1980). The feature is most typical of the Phyllostomidae, but occurs sporadically in other groups.

Students of bat communication should certainly take into account the variable development of the vomeronasal organ complex, but data involving this feature with communication behavior are conspicuous by their absence. The focus in terms of explanation thus far has been on diet (Bhatnagar 1980).

Source of Olfactory Signals

As clearly pointed out by Eisenberg and Kleiman (1972), in a receiver system as sensitive as the olfactory assemblage in mammals, the whole corpus of the mammalian sender can generate a wealth of signals. Urine, feces, saliva, and the products of specific glands may be of particular importance. The functional significance of the different products is not always clearly identified for mammals in general (Eisenberg and Kleiman 1972) and bats in particular.

Although urine and feces often conspicuously contribute to the atmosphere in the roosts of bats, there appears to have been no systematic study of their role in communication between bats. Buchler (1980b) found evidence of a "scent post" in *Myotis lucifugus*, but was unable to categorically identify urine or feces as the important active component.

Bats possess a wide range of integumentary glands (Quay 1970), most of which are undescribed anatomically, let alone in any behavioral context. There are sudoriferous glands (coiled tubular structures opening into hair follicles), either apocrine or merocrine glands producing odorous lipids and/or sweat. These are under nervous and blood hormone control (Muller-Schwarze 1983). There are also sebaceous (= holocrine or alveolar) glands producing lipids by the disintegration of cells, and whose action is controlled by blood hormones, usually steroids (Stoddart 1980).

Table 6 presents an outline of the distribution of different glands among different families of bats. The important conclusion to be drawn from these data is that bats possess an impressive array of glands whose products may serve important roles in communication. In many cases, an important role in sexual behavior is implied by strong sexual dimorphism in glandular structure or appearance (e.g., Mainoya and

TABLE 6. Distribution of scent glands among different families of bats (after Quay 1970)

Family	PROPATAGIAL	UROPATAGIAL	WRIST AND ANKLE	CHEST	CIRCUMANAL	POSTANAL	PARAANAL	PREANAL	AXILLARY	SHOULDER	BACK	GULAR	NECK	THROAT	LIPS	CHIN	PARARHINAL	FACIAL	EARS	INTERAURAL
Pteropodidae						X	♂			♂			♂							
with tuft										X			X						X	
Rhinopomatidae																	X			
Craseonycteridae																				
Emballonuridae	♂	X		X								♂		♂	X	♂				
Noctilionidae																				
Megadermatidae											♂									
Nycteridae									X									X		
Rhinolophidae																		X		
Hipposideridae								X										X		
Mormoopidae																				
Phyllostomatidae										X		X	X		X			X		
Thyropteridae																				
Furipteridae																				
Natalidae																	X			
Vespertilionidae					X							X			X		X			
Mystacinidae																				
Myzopodidae			X																	
Molossidae												X								X
with tuft																				X

X - present in male or female; ♂ males only

Howell 1977). There is considerable evidence of seasonal changes in the glands of some bats (e.g., Dapson et al. 1977), sometimes in association with lactation (e.g., *Eptesicus serotinus,* Kleiman 1969; *Nycticeius humeralis,* Watkins and Shump 1981).

Only further study will clarify the roles of the products of these glands in the communicative behavior of bats. Some behavioral and ecological correlates of mammalian scent marking are provided by Muller-Schwarze (1983) and Bronson (1983), including and lacking, respectively, any information about bats.

AUDITORY SYSTEM

The auditory system of bats is, for the most part, typically mammalian, and is reviewed in detail in Henson (1970a). In many species the pinnae are large and conspicuous, particularly in the Microchiroptera. Sound waves are converted to mechanical vibrations by the tympanum and these vibrations are amplified by the auditory ossicles. At the oval window the mechanical vibrations are converted to vibrations in fluid, specifically pressure waves in the cochlea. Tips of hair cells embedded in the tectorial membrane constitute the basilar membrane, which converts vibration to neural signals, presumably by changes in permeability. The narrow, basal end of the basilar membrane is sensitive to high-frequency sounds, the broader, distal end to lower frequencies. Differences in the structure of the basilar membrane act to tune the auditory systems of some bats in connection with neurological tuning. There are also variations in the terminal zone of the external auditory meatus, which are not yet fully understood (Di Maio and Tonndorf 1978).

The auditory nerve conducts stimuli to the brain in typically mammalian fashion. The ascending pathway involves the cochlear nuclei, the trapezoid bodies, the superior olive, the periolivary nuclei, the lateral lemniscus, the inferior colliculus, and the median geniculate body before information is relayed to the auditory cortex (Neuweiler 1980a). Echolocating bats may have a unique feature, namely the intermediate nucleus of the lateral lemniscus, receiving input from the anteroventral cochlear nucleus and sending major projections to the central nucleus of the inferior colliculus (Zook and Casseday 1980). There appears to be point-to-point representation between the organ of corti and the auditory cortex.

In general, the sensitivity of bat hearing to high-frequency sounds is well established by a variety of audiogram techniques (Novick 1977). In species with specialized basilar membranes and associated neurological adaptations, frequency resolutions in the order of 10 Hz may be achieved (Neuweiler 1980b). Bats also possess exceptional abilities for making fine discriminations in time, appropriate for animals which echolocate (Simmons and Stein 1980).

It is now clear that bats are also sensitive to sounds of lower frequency (Fig. 13), as indicated by a range of behavioral evidence (e.g., Fiedler 1979; Tuttle and Ryan 1981; Bell 1982a). Many species use low-frequency sounds emanating from prey, associated with wingbeats (Bell 1982a, 1982b), rustling sounds (Fiedler 1979), or mating calls (Tuttle and Ryan 1981), to find their food. It is clear that further work will reveal sensitivities to lower-frequency sounds not explored by some of the original studies.

The vocalizations of bats cover a wide range of frequencies, from those clearly audible to unaided human ears, to others which may reach over 200 kHz. As we shall see, it is not reasonable to arbitrarily divide bat vocalizations into those functioning in echolocation versus those used in communication, since many in the former category serve a communication function. In the same way, echolocation does not always involve ultrasonic (beyond 20 kHz) calls.

Auditory systems, from sound reception to information processing, are highly developed in bats, particularly in the echolocating forms, which show greater development of acoustical centers in the brain than do non-echolocating species (Henson 1970b).

Sound Production

Bats may produce sounds by a variety of mechanical means typical of many mammals, but also including sounds associated with wingbeats. Many bat sounds are vocalizations emanating from the larynx, which is the source of echolocation sounds of the Microchiroptera (Novick 1977). The larynx of many Microchiroptera, and particularly vespertilionids, as suggested by Suthers and Fattu (1982), is adapted for production of brief ultrasonic pulses of high intensity and at a high repetition rate. The vibrating elements are the vocal membranes and perhaps the ventricular folds, which are surrounded by a massive framework of muscular and cartilagenous material. Tension of the vocal membranes is controlled by contraction and relaxation of the

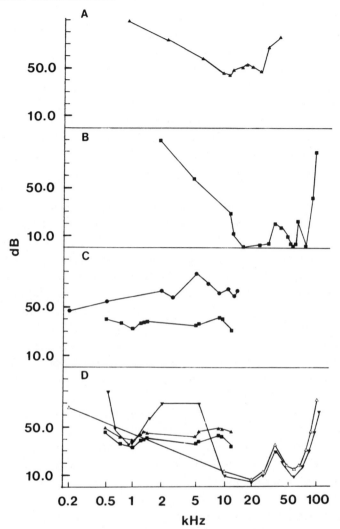

FIG. 13. A comparison of behavioral audiograms for several species of bats. Shown here are data for the echolocating pteropodid *Rousettus aegyptiacus* A (data from Suthers and Summers 1980); the Doppler-shift compensating *Rhinolophus ferrumequinum* B (data from Long and Schnitzler 1975); the phyllostomids *Trachops cirrhosus* (●) and *Macrotus californicus* (■) C (data from Ryan, Tuttle, and Barclay 1983, and V. Cuccaro, respectively); and the vespertilionids *Eptesicus fuscus* (Δ from Dalland 1965; ▲ from V. Cuccaro, pers. comm.; ▼ from Poussin and Simmons 1982), and *Antrozous pallidus* (■ from V. Cuccaro, pers. comm.) D. Note that some species, particularly the phyllostomids and vesperilionids for which there are data, are sensitive to lower-frequency sounds.

cricothyroid muscles. The innervation of this system is similar to that of other mammals (Suthers and Fattu 1982).

Because of their low intrinsic mass, the vocal membranes of vespertilionids (and perhaps many other microchiropterans) vibrate at ultrasonic frequencies. Phonation is preceded by closing of the glottis. Initiating expiratory effort raises the subglottal pressure, which, with the relaxation of the glottal adductor muscles, slightly opens the glottis and permits airflow past the vibrating membranes. Frequency of the vibrations is determined by tension on the membranes, and bats control their vocalizations accordingly. The cricothyroid and other intrinsic laryngeal musculature control the intensity (sound pressure level) of the emitted sounds (Suthers and Fattu 1982). In general, a glottal gate is central to phonation in some, if not most, echolocating bats (Suthers and Fattu 1982). An additional adaptation associated with the production of high-intensity pulses is the static lung compliance. Bats have stiffer lungs than do many other mammals, and less compliance may increase recoil pressure and the driving force in phonation. A noncompliant lung may prevent early airway closure (Suthers and Fattu 1982). The ventral surface of the larynx is covered by a large elastic cricothyroid membrane, which could act as a compliant reservoir of subglottal air (Fattu and Suthers 1981). It is important to note that even during inspiration, phonation in some bats is accompanied by brief exhalation (Suthers and Fattu 1982).

This detailed information about phonation is not available for most species of bats, so it is likely that variations on the theme will be discovered when more work is done. Echolocating megachiropterans, some species in the genus *Rousettus*, produce their orientation calls by clicking their tongues against their palates (Roberts 1975), a situation easily demonstrated by severing the hypoglossal nerve (Novick 1977). Other studies have examined the auditory sensitivity of *Rousettus aegyptiacus* (Belknap and Suthers 1982), the species on which most work has been done.

In addition to vocalizations associated with orientation or echolocation, bats produce a wide range of sounds important for communication. There is a general trend for species with more complex social organization to have more elaborate repertoires, but the data base currently is small compared to the diversity of bats.

III

Bat Communication

ECHOLOCATION SIGNALS IN COMMUNICATION

In a review of bat social organization and communication, Bradbury (1977a) pointed out that we must eventually come to grips with the question of how much bats rely on high-frequency sounds for communication. An obvious extension of this, noted by Gould (1971), Brown (1976), and Bradbury (1977a), among others, is an examination of the extent to which individuals use echolocation signals for communication. Inherent in this situation is the often-assumed generalization that bats tend to use lower-frequency signals for social interactions and higher-frequency ones for echolocation. It is obvious that many bats, for example the emballonurid *Saccopteryx bilineata*, mix social signals with biosonar pulses, and in many instances adjacent emissions involve drastically different frequency spectra. A wide range of frequencies can be achieved by the deletion or addition of harmonics (Bradbury and Emmons 1974). Echolocating bats adjust the frequencies of their vocalizations by altering the tension on the vocal membranes. There are similar reports of a wide dynamic range of bat vocalizations for several species (e.g., Fenton 1980; L. A. Miller and Degn 1981; Swift 1981).

It is commonly accepted that calls used in echolocation by bats have precursors in communication, and that echolocation could have evolved through use of vocalizations produced in another, presumably

63

social, context (e.g., Ewer 1968; Gould 1971). The echolocation calls of some swiftlets (*Collocalia* spp.) also have communication precursors (Suthers and Hector 1982). Gould (1983) reviewed the importance of rhythmic patterns with signal perception and production in a variety of situations where the patterns provide a pacemaker for sniffing, chewing, and licking. He noted that a "put" sound is an acoustic consequence of active olfactory exploration, and permits sensory input via olfactory, acoustic, and tactile (vibrissae) modes. Komisaruk (1977) suggested that this pacemaker, located in the lower brainstem, involves "theta waves," and Gould (1983) pointed out that this situation of coordination of sensory input during exploration "is consistent with the hypothesis that bat sonar evolved from continuous, graded communication signals of early insectivores."

Möhres (1967) addressed the question of echolocation and communication by observing a captive colony of the rhinolophid *Rhinolophus ferrumequinum*. He suggested that these bats used the echolocation calls of roost-mates to obtain positional information about preferred roost-mates, and his patterns of observation and manipulation made it clear that the bats used pulses normally associated with echolocation in a communication role.

Common sense dictates that the vocalizations one animal uses for echolocation are available to other animals within earshot, implying, of course, that the calls are audible. Since many bats use high-intensity calls for echolocation, they may be clearly audible over a relatively large area. Furthermore, echolocation pulses are often produced at high repetition rates, from 50 to over 500 sec^{-1}. In short, the echolocating animal gives away a great deal of information about itself, its position and course. Since many species of echolocating bats produce calls for orientation which are species-specific (in frequency-change over time; Fenton and Bell 1981; Fenton 1982b), given the capacity of many Microchiroptera for analysis of sounds, the potential information available to an eavesdropper is considerable (Fig. 14).

Another piece of evidence supporting the role of calls presumably designed for echolocation in a communication sense is provided by changes observed in design of echolocation calls according to the presence or absence of conspecifics. The most elegant demonstration of this to date is the work of Habersetzer (1981) on the rhinopomatid *Rhinopoma hardwickei* in India. When these bats were returning to their roosts or preparing to depart from them, the bats used distinctive steep

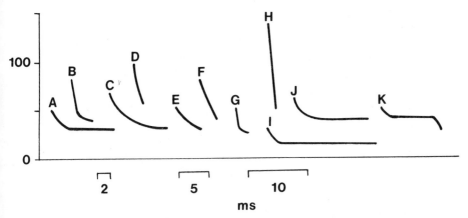

FIG. 14. Different species in a fauna of insectivorous bats produce diagnostically different echolocation calls. Shown here are the calls of the bat community around Portal, Arizona, including *Lasiurus cinereus* (A), *Myotis volans* (B), *Eptesicus fuscus* (C), *Pipistrellus hesperus* (D), *Myotis thysanodes* (E), *M. californicus* (F), *Antrozous pallidus* (G), *Myotis auriculus* (H), *Tadarida macrotis* (I), and *Tadarida brasiliensis* (J—echolocation call; K—honk, see also Fig. 18). Modified from Fenton and Bell 1981.

FM signals, while other signals, presumably designed for communication, were produced by roosting bats. When several *R. hardwickei* were flying in the open they used CF signals, with the frequencies different between different individuals. Solitary *R. hardwickei* foraging in the open produced CF calls at one frequency. Habersetzer (1981) suggested that modification of the CF frequency according to company might permit the bats to minimize jamming interference. The important point of Habersetzer's observations from a communication standpoint is the way the echolocation calls were adjusted to the social setting. (Fig. 15.)

There are other reports of echolocating bats altering the frequencies of their calls depending upon the situation, presumably not with respect to target orientation. Roberts (1972) found changes in the CF frequencies of echolocating bats in east Africa. Barclay (1983) made some detailed studies of a variety of emballonurids in Panama, elaborating on earlier work on some of these species by Pye (1978). Barclay found that individual bats changed the frequencies of their echolocation emissions rather dramatically from pulse to pulse over a range of two or three patterns (Fig. 16). In his observations, there was no clear

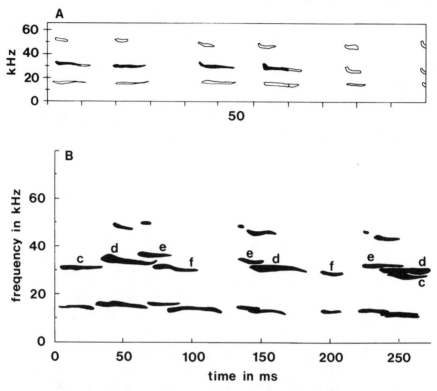

FIG. 15. The ultrasonic echolocation calls of a single *Rhinopoma hardwickei* (A) flying in the open, compared to those from a group of four bats (B) flying in the open. Note that the four individuals (c, d, e, and f) emit different frequencies best seen in the second harmonic (labelled). Modified from Habersetzer 1981.

change in the social setting in which the calls were produced, but the changes in the calls could have reflected a concurrent role in communication.

The first experimental demonstration of a communication role for echolocation calls involving playback presentations of vocalizations is the work of Barclay (1982a). He showed, by presentations to free-flying *Myotis lucifugus* in the field, that individuals used the echolocation calls of conspecifics to locate clumped resources, including food, roosts, and mating and hibernation sites. His playback experiments included controls designed to challenge the idea that the bats were merely curious about any appropriate sounds, and he found that young of the year were more responsive to playback presentations of echolocation calls than were adults. The behavioral response he observed involved flights

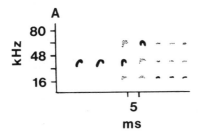

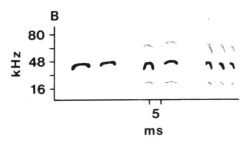

FIG. 16. Variations in the echolocation calls of two neotropical emballonurids, *Cormura brevirostris* (A) and *Saccopteryx bilineata* (B), adapted from Barclay (1983). Note changes in frequencies of harmonics with most energy.

toward the speakers from which the stimuli were presented; responses were lacking or significantly reduced during control presentations. Barclay (1982a) noted that a hunting *M. lucifugus* could detect a hunting conspecific at a range of up to 50 m; the same bat probably detects a 15 mm diameter target at a range of 2 or 3 m, if the data of Kick (1982) are representative. Careful analysis of the tapes Barclay (1982a) used in presentation experiments indicated no evidence of anything other than "normal" echolocation calls (Fig. 17).

Further experiments, by Leonard and Fenton (1984), demonstrated that responses to echolocation calls are not limited to *M. lucifugus* and that responses are not uniform between species. Leonard and Fenton used playback presentations to free-flying *Euderma maculatum* (Vespertilionidae) in the field to investigate the role of echolocation calls in the spacing of these bats. *Euderma maculatum* normally hunt alone, and one rarely sees two bats within 50 m of one another except when they are emerging from or returning to their roosts. The presentations, again including controls, produced one of three responses. The bats paid little attention to control stimuli, but always responded to presentations

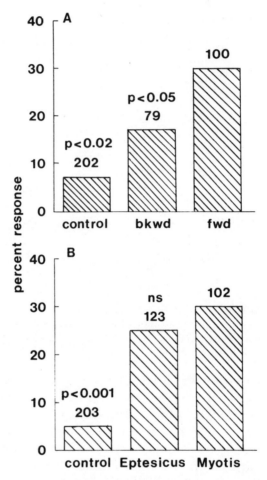

FIG. 17. Responses of free-flying *Myotis lucifugus* to various stimuli: A—response to recordings of feeding *M. lucifugus* played forward (fwd) and backward (bkwd), as compared to the response to silent controls; B—comparisons of the response levels to playbacks of *M. lucifugus* versus the echolocation calls of *Eptesicus fuscus* and silent controls (adapted from Barclay 1982a).

of the echolocation calls of conspecifics. In a playback situation, *E. maculatum* either flew toward the speaker and produced a vocalization (an "irritation buzz"), or turned and left the immediate area. In this species, individuals appear to rely on echolocation calls of conspecifics to achieve spacing of the population on feeding grounds.

These two experiments show that the vocalizations one animal uses to gather information about its surroundings can simultaneously serve additional functions. It is obviously inappropriate to classify the vocalizations of bats as useful either in echolocation or in communication. Some ecological studies have implied that different bats cued on the calls of one another, and this cuing sometimes involved more than one species (e.g., Fenton and Morris 1976; Bell 1980a). Barclay's (1982a) presentations indicated that *Myotis lucifugus* would respond to the echolocation calls of another species, *Eptesicus fuscus,* and vice versa.

There is obviously much work to be done on the role of echolocation calls in communication. Thomas, Fenton, and Barclay (1979) found that vocalizations were important in mating interactions between male and female *Myotis lucifugus,* but that the vocalizations were often those associated with echolocation. There are similar trends in the interactions between mother bats and their young.

Suthers (1965) demonstrated that an echolocation call could be slightly modified to enhance its communication function. He found that when two noctilionids *(Noctilio leporinus)* were on a collision course, one would "honk" at the other, resulting in one of the bats veering away and avoiding a collision. Honking involved dropping the frequency sweep of the echolocation call an extra octave (Fig. 18), and similar modifications to echolocation calls are now known from some other echolocating bats (e.g., Fenton and Bell 1979; 1981; Fenton 1980).

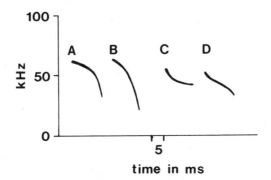

Fig. 18. Under some circumstances when two bats are on a collision course, one or the other or both will adjust their echolocation calls by lowering the terminal frequency and producing a "honk." Compared here are the regular echolocation calls (A and C) and honks (B and D) of *Noctilio leporinus* (A and B) and *Myotis volans* (C and D). Modified from Suthers (1965) and Fenton and Bell (1979).

There is clearly a bridge across the range of vocalizations used by bats. At one end are calls useful in echolocation but exploitable in communication; in the middle are echolocation calls modified to enhance a communication function; at the other end are vocalizations presumably functioning only in communication. Experimental evidence is required to confirm that "social calls" do not serve any function in echolocation.

FEEDING

There are not many observations of interactions between feeding bats, and few permitting categorical identification of the mode of communication which led to the interaction. In many cases interactions arise from sounds associated with feeding and not intended as displays. For example, in a study of captive *Nycteris grandis,* Fenton, Gaudet, and Leonard (1983) found that the chewing sounds of one bat attracted the attention of others. Furthermore, they reported that an apparently dominant individual, having finished her first prey item, tried to steal food from her roost-mates. As a result, when the dominant individual stopped chewing, the other five bats also stopped chewing and would not resume eating until the dominant bat had started to eat again.

In an effort to train bats to perform various tasks associated with finding food, Gaudet (1982) discovered that bats quickly associated chewing sounds with feeding. She demonstrated that chewing, even by human observers, attracted the attention of hungry bats. In some species, for example *Myotis lucifugus,* altercations over food were relatively uncommon in a captive situation, while other species, notably *Eptesicus fuscus,* were highly aggressive, particularly over limited supplies of food. *Antrozous pallidus* usually showed a low level of aggression around food (Gaudet 1982), but under natural situations bats removing prey from the ground and taking it to a roost for consumption usually did so secretly and, when roosting, enveloped their prey with their wings and avoided contact with roost-mates (Bell 1982a; pers. comm.).

It is clear that sounds associated with food and feeding elicit a great deal of attention from some bats. Much of our knowledge of these interactions derives from studies performed in captivity, a logical outcome of the often dispersed nature of many of the food resources exploited by bats.

The different food-consumption habits of some fruit-eating bats may be significant. Some species typically land at a food site, quickly select a fruit, and take it to a night roost for consumption (e.g., many phyllostomids, Fleming 1982). Other species, notably some of the larger pteropodids, such as *Pteropus* spp. and *Eidolon*, remain at the source of the food and attempt to defend it from other bats (Rosevear 1965; Nelson 1965; Gould 1977b). Altercations at the sources of food often involve loud vocalizations, making the feeding sites of some species conspicuous by the background of bats bickering over food and/or space (e.g., Rosevear 1965).

Removal of food for consumption at night roosts is often interpreted as an antipredator strategy (Fleming 1982). But it could be an effective means of avoiding or minimizing altercations with other bats at the food source and so facilitate undisturbed feeding. Obviously the conditions leading to night-roosting behavior will not be common to all bats. Species such as *Myotis lucifugus* use night roosts not for consumption of food but for thermoregulatory advantages (Anthony, Stack, and Kunz 1981; Barclay 1982b). Barclay (1982b) has shown recently that the patterns of arrival at and departure from night roosts are not random, and suggested that night roosts could serve as information centers, an idea also proposed for a variety of communally roosting birds. Barclay's (1982b) study did not indicate just how information might be transmitted from one individual to another within the night roost, but the possibility remains.

Data on the feeding behavior of nectar- and pollen-feeding bats are limited. D. J. Howell (1979) showed that the phyllostomid *Leptonycteris sanborni* forages in flocks, which provides the bats with benefits in food location and temperature regulation. The low-intensity echolocation calls of these bats make their acoustical communication difficult to study, and at this time it is not clear if the feeding behavior of *L. sanborni*, a relatively specialized nectar-feeder relying on pollen as a protein source, is typical of nectar-feeding bats in general, or even phyllostomids in particular. Other work by D. J. Howell (1974) indicates that not all of the nectar-feeding phyllostomids she studied are as specialized as *L. sanborni*.

For animal-feeding bats, data on communication about food resources are relatively sparse. There is evidence of territoriality in the emballonurid *Saccopteryx leptura*, which actively defend feeding areas and roost sites (Bradbury and Emmons 1974). The echolocation calls of hunting bats provide cues about their presence and activities, to intruders or to territorial males (Bradbury and Emmons 1974). Vocalizations other than echolocation calls coincide with altercations associated with violations of territorial boundaries, but a detailed analysis of the vocalizations involved has not been presented. It is important to note that altercations involving vocalizations which are audible to man are more conspicuous than are ultrasonic interactions (see below).

Vaughan (1976) reported evidence for exclusive use of feeding areas by the African megadermatid *Cardioderma cor*. These bats usually forage from perches close to the ground, commonly pursuing prey moving along the ground, but often chasing flying prey, sometimes while flying continuously themselves. *Cardioderma cor* produce low-intensity echolocation calls, which precluded Vaughan's (1976) establishing whether or not they used echolocation as they hunted.

Most striking, however, was the "singing" behavior of *C. cor* as it moved about its feeding area. Singing was strongly seasonal in occurrence (Fig. 19), and this pattern of behavior influenced the success of the bats during times when food was less abundant (Vaughan 1976). It is tempting to speculate that reliance on low-intensity echolocation calls necessitated the use of the louder, lower-frequency vocalizations to advertise the presence of the bats. This situation offers an excellent opportunity for playback experiments.

The system which Vaughan (1976) studied implied mutual avoidance as opposed to territorial interactions, since he provided no evidence of defense. The work on *Saccopteryx leptura* had involved feeding territories, since the sites were actively defended (Bradbury and Emmons 1974). Wickler and Uhrig (1969) used two weeks of observation of the African megadermatid *Lavia frons* to suggest the use of feeding and roosting territories.

Altercations between flying, feeding bats have often been reported, but they may not always be what they seem. Bell (1980b), in a study of habitat use by bats in Arizona, observed a sequence of altercations between a vespertilionid, *Lasiurus cinereus,* and three other bats, two vespertilionids (*Eptesicus fuscus* and *Lasionycteris noctivagans*) and a molossid *(Tadarida brasiliensis).* One interpretation of these interactions would invoke a pattern of use of space and territoriality. The interactions could also have been interpreted as evidence of carnivory by *L. cinereus.* The interactions, however, were probably due to the fact that the *L. cinereus* was afflicted with furious rabies. In the New World,

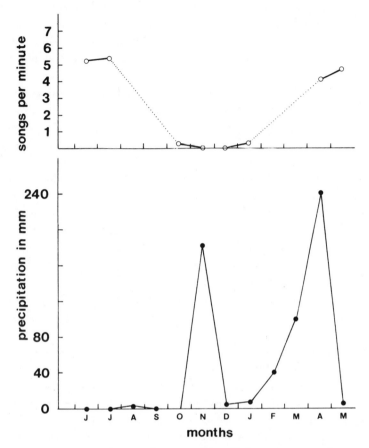

FIG. 19. The incidence of songs of *Cardioderma cor* per minute (above) during the period June 1973 to May 1974 is compared to the monthly pattern of precipitation (below). Dotted lines indicate times when no accurate counts of songs were made; note that there was little singing during the November rains and that singing reached its peak during the March–April rains. In this species the songs (which are audible to the unaided human ear) appear to serve a spacing function. Modified from Vaughan (1976).

behavioral changes associated with rabies remain a possibility when observing interactions between bats in the field.

I have already reviewed the work of Barclay (1982a) and Leonard and Fenton (1984), which indicated that feeding bats may attend to the echolocation calls of others, usually conspecifics. Responses permit them to cue in on concentrations of food, or to avoid conspecifics in the field. Also noted was honking behavior reported from bats on collision courses. There is obviously communication between feeding insectivorous bats in the field involving modified or unmodified echolocation calls.

There are other observations of interactions between bats presumed to be feeding, usually conspicuous because they involve vocalizations audible to the unaided human ear. In some cases, bats alternate production of social calls with echolocation calls. Miller and Degn (1981) reported that three species of vespertilionids commonly alternated typical echolocation calls with vocalizations presumed to have a social function. In two of these species, *Nyctalus noctula* and *Eptesicus serotinus*, the combined exchanges of vocalizations occurred in the vicinity of roosts during the mating season, and there was no strong evidence that the bats producing the calls were actively foraging. The third species, however, *Pipistrellus pipistrellus*, was clearly feeding at the time the altercations were observed. The bats that Miller and Degn (1981) recorded modified different parts of their echolocation calls according to the company they kept, and included other vocalizations in the altercations.

More light is shed on the situation involving *P. pipistrellus*, by the work of Swift (1981) in Scotland. She observed altercations between feeding *P. pipistrellus* at study sites and demonstrated a strong positive correlation between the incidence of the altercations and the availability of insect prey. When insects were abundant, the incidence of interactions was very low. Unfortunately, her study did not document possible ultrasonic (and therefore inconspicuous) interactions.

Some male *Pipistrellus nanus* chased other males from 24 m long elliptical areas when food was abundant, but as the dry season progressed, they spent more time feeding away from the originally defended areas (O'Shea 1980). O'Shea found that on other occasions this species foraged in groups in other situations. Altercations over the use of foraging space did not involve vocalizations audible to O'Shea, although advertisements by males at roosts did.

In Hawaii the vespertilionid *Lasiurus cinereus* shows a high level of aggressive interactions in feeding areas, including chases and vocalizations audible to man; the incidences of aggressive interactions appeared to be associated with abundance of prey (Belwood 1982; Fullard 1982b). In Manitoba, the same species defends patches of prey only when the prey is highly clumped, a situation associated with prevailing winds (Barclay 1982c). Other workers have suggested that *Lasiurus borealis* is territorial (Barbour and Davis 1969), but variations in this pattern, including some clear evidence of multiple use of some resources by many bats, confound the situation. The distribution of prey may strongly influence aggressive levels among feeding bats.

Wallin (1961) reported altercations between foraging *Myotis daubentoni* hunting over a marshy area, while Miller and Degn (1981) found no evidence of such behavior in their work on this species. Tuttle (pers. comm.) observed territorial encounters between feeding *Myotis grisescens* under some conditions of prey distribution, but not under others; he did not report whether or not vocalizations audible to him accompanied these territorial encounters.

For many other species, studied in the field to varying degrees, there is no evidence of territorial interactions. For example, *Myotis lucifugus* has been studied in the field in considerable detail by several workers and none has reported evidence of aggressive encounters between foraging individuals (e.g., Fenton and Bell 1979; Anthony and Kunz 1977; Harrison 1983). Fenton and Bell (1979) found no evidence of aggressive encounters between foraging *Myotis volans, M. californicus,* or *M. auriculus,* and in some bat communities such encounters are rare (Bell 1980a). Although some pairs of *Myotis* spp. show preferences for some habitats over others, e.g., *Myotis californicus* and *M. leibii, M. lucifugus* and *M. yumanensis,* the habitat partitioning does not involve aggressive interactions (Woodsworth 1981, and Herd and Fenton 1983, respectively).

It is important to remember that aggressive interactions involving vocalizations audible to the observer are conspicuous, and it is possible that some of the studies cited above which did not monitor ultrasonic vocalizations overlooked aggressive encounters. Clearly the echolocation calls could by themselves constitute advertisement of preferred access to feeding areas. I have reported some observations of foraging *Eptesicus fuscus* that clearly involved alternations of social and echolocation calls. In that setting, playback of echolocation calls of this species recorded some distance away increased the incidence of interactions, suggesting that the aggressive bat(s) monitored the echolocation calls of others foraging in the area (Fenton 1980).

There remains the distinction between signals not intended for communication, and therefore not strictly communication displays, and those designed to transmit information. In the case of echolocating animals hunting prey, the information is available while the animal is foraging, suggesting an important reason for the bats to avoid echolocating in some circumstances. Even in the absence of echolocation, the successful forager, the one chewing its food, also divulges information about itself. This situation, as we have seen, could influence patterns of food consumption for different bats.

It is clear that some species are predisposed to respond to certain stimuli, typically, for example, those stimuli associated with feeding by a conspecific (Galef 1976). Sensitivity to stimuli associated with feeding has been demonstrated to influence foraging behavior. Gaudet and Fenton (1984) trained individual *Myotis lucifugus, Eptesicus fuscus,* and *Antrozous pallidus* and then provided naïve conspecifics recently captured in the field with the chance to observe the trained bats as they fed. Bats exposed to feeding conspecifics learned the new feeding behavior significantly faster than a control group learned by manual conditioning (Fig. 20); another control, provided with the chance to learn by trial and error, never acquired the new feeding behavior (Gaudet and Fenton 1984). This work also demonstrated that learning by observation could occur between species, but in that situation the learning was strongly influenced by levels of aggression.

Gaudet and Fenton (1984) provided the bats playing the roles of "pupils" with the chance to interact freely with the conditioned bats, exposing them to a range of stimuli, including echolocation sounds, chewing sounds, and other patterns of behavior. Since the three species studied show a spectrum of social organization from gregarious (*M. lucifugus* and *E. fuscus*) to more social (*A. pallidus*), observational learning is obviously not restricted to the more social bats.

These results bear directly on the matter of communication between individuals, whether intentional or not. They have important implications for studies of foraging behavior and ecology, and may prove particularly relevant to investigations of interactions between mother bats and their offspring. Vaughan (1976) reported that young *Cardioderma cor* accompany their mothers during foraging, and similar patterns may occur in species with long periods of weaning (e.g., *A. pallidus,* Brown 1976; *Desmodus rotundus,* Greenhall, Joermann, and Schmidt 1983).

There is tremendous potential for communication between foraging and feeding bats.

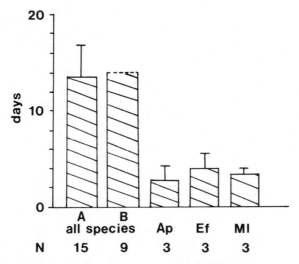

Fig. 20. The number of days required for bats to learn a food location task provides evidence of observational learning in *Antrozous pallidus* (Ap), *Eptesicus fuscus* (Ef), and *Myotis lucifugus* (Ml). The conditioned controls (A) took significantly longer than the bats learning through observation (Ap, Ef, Ml), while trial-and-error controls (B) never did learn the new feeding behavior and the trials were terminated after two weeks. There were no significant differences between the performances of different species (bars are standard deviations of the mean); N = numbers of individual bats. Modified from Gaudet and Fenton (1984).

ROOSTS

Many species of bats roost alone except when a female is caring for her current young. Others are gregarious, often roosting in large numbers in close proximity to or in contact with other bats, usually of the same species. In many roosting aggregations, there is no obvious social structure. Still other species roost in highly structured social settings, situations where individuals aggregate nonrandomly with roost-mates. The level of communication associated with roosting reflects the setting, increasing in complexity with increased social structure. Seasonal changes in behavior complicate the picture, since structured social populations may occur only during the mating season, which is often quite divorced in time from parturition. Differences in social status also influence roost aggregations (Fig. 21).

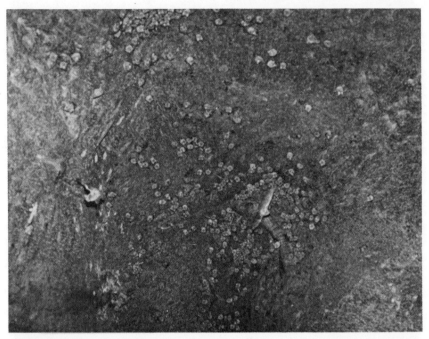

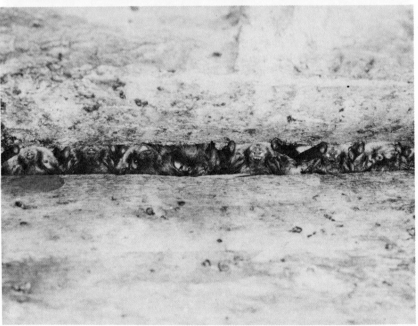

FIG. 21. Photographs showing roosting aggregations of bats. The ceiling of the hollow in a baobab in Zimbabwe occupied by roosting *Rhinolophus hildebrandti* (A) clearly indicates that while some individuals roost in close contact with one another, others are solitary. When this photograph was taken the bats were not in mating condition and young were almost full-grown. In a maternity roost, female *Myotis lucifugus* (B) squeeze together in a long narrow crack. These animals are active and pregnant or lactating; this species will also roost in more open hollows. Hibernating *Myotis sodalis* (C) cluster on the ceiling of a cave in close physical contact with one another. The function of clustering by hibernating bats is not clear, but in many species is not relatable to thermoregulation. B and C reproduced with permission of the University of Toronto Press.

In many cases data on activity within roosts are limited and mislead-
ing, reflecting disturbance by the observer. Roosts housing large numbers of bats are often uncomfortable for people, and prolonged
observation of marked animals, essential to clarification of interactions
among roost-mates, are scarce. In spite of these caveats, however, it is
possible to assemble some generalizations about communication between roosting bats.

Species which appear to be solitary have been little-studied in their
roosts. For example, the vespertilionid lasiurines of the New World are
presumed to roost alone in foliage, although one species, *Lasiurus
intermedius*, forms nursery colonies among dried foliage in some situations (Barbour and Davis 1969). Some workers have the impression that
lasiurines such as *Lasiurus borealis* are territorial in feeding areas, so
that solitary roosting could reflect other patterns of spatial use. Barclay
(1982c) found that in some areas individual female *Lasiurus cinereus*
occupy the same roosts during the period when they are caring for
their young and for some time after the young are weaned, but other
records (Constantine 1966; Downes 1964) indicate successive use of
roosts by different individuals, sometimes of different species. That
different bats sequentially occupy the same roost site probably reflects
availability of appropriate roost sites, but could involve some unde-
tected marking of sites by the bats. Further observation is required to
clarify the situation.

The data make it difficult to determine if solitary roosting is the
result of mutual avoidance, spacing of appropriate roosts, or some
active defense or advertising of roosting sites. Since *L. cinereus* and *L.
borealis* select particular roost sites, well protected from the top and
sides (Constantine 1966), and since these conditions are presumed to
apply to other lasiurines (Barbour and Davis 1969), a shortage of
suitable roosts could influence the dispersion of the bats. That *L.
intermedius* forms nursery colonies only confounds the issue.

I suspect that many other species of bats have roosting patterns
similar to those described for the lasiurines, and in some cases these
may be species known from very few specimens (Walker et al. 1975).
The lack of observations of these cryptic animals makes it difficult to
put their patterns of behavior into any logical context with respect to
communication. The most parsimonious interpretation of solitary
roosting is random occupation of acceptable roost sites.

Other species, some of the best examples coming from the Pteropodidae, form roosting assemblages ranging from a few individuals (e.g., *Epomophorus gambianus*) to large, conspicuous aggregations (e.g., *Eidolon helvum* and many species of *Pteropus*). There are observations of roosting interactions among some of these bats, but for other pteropodids, which apparently roost alone (e.g., *Micropteropus pusillus* or *Nanonycteris veldkampi*), there is no information about their roosting behavior.

Marshall and McWilliam (1982) have provided some observations of roosting behavior of *Epomophorus gambianus* in Ghana, where these bats form colonies in tall trees. A colonial roost involves individual bats that are physically isolated from their neighbors, and in these settings they remain relatively quiet during the day, showing little movement or grooming. The roosting bats are, however, alert, and a close approach by a neighbor elicits a "quarrelsome call" from the bat whose space has been intruded into. Similar spacing of individuals is known from other epomophorine bats such as *Hypsignathus monstrosus* (Bradbury 1977b), and *Epomophorus wahlbergi* (Wickler and Seibt 1976). Colonies or aggregations may involve anywhere from five to ten or over fifty bats, but the roosting assemblages are cryptic and the location of a colony is not immediately obvious to the casual observer. Spacing within the roosts is maintained by each animal protecting its individual space or by mutual avoidance. In some *Epomophorus* spp. changes in posture and vocalizations are commonly used to warn intruders or neighbors against a close approach.

Other colonial pteropodids show similar patterns of spacing of individuals maintained by aggressive vocalizations and visual (postural) displays, but some other colonies are not cryptic. Nelson found that territorial male *Pteropus poliocephalus* marked their roosts with exudate from the scapular glands by rubbing the gland along the branches on which the bat roosted. These bats also challenged intruders with a territorial warning call. There are, however, at least two important differences between aggregations of epomophorine bats and those of *Pteropus* spp. One is the level of activity in the roost, for in their camps the *Pteropus* are more active and noisy, and this, combined with the larger populations in the colonies, makes their roosts easy to find. The second is that neighboring *Pteropus* (Nelson 1965; Neuweiler 1969) show great interest in one another and often engage in mutual grooming. The territorial behavior is short-lived. *Eidolon helvum* form large camps, and within these camps the high level of activity, including

movement to alternate perches and bickering, makes the sites conspicuous (Rosevear 1965; Kingdon 1974). Unfortunately details of individual behavior in roosts are lacking.

In some seasons the camps of *P. poliocephalus* are structured, showing
varying degrees of sexual segregation or with younger animals rele-
gated to more peripheral positions (Nelson 1965). Individual spacing is
maintained by appropriate communication signals, from vocalizations
to scent marks and visual displays, but although individuals may often
roost alone, they show high levels of interaction with and interest in
conspecific neighbors. Roosts within foliage provide the maximum
opportunity for use of the full range of communication media, given
the levels of ambient lighting and the availability of surfaces for scent-
marking.

In contrast to the pteropodids, many of the Microchiroptera roost in
close proximity to, if not in physical contact with, their immediate
neighbors. For example, in their tents, *Ectophylla alba,* a phyllostomid,
are in physical contact with one another during the day (Timm and
Mortimer 1976). In some situations, the level of proximity and contact
reflects the ambient temperature conditions. Licht and Leitner (1967)
demonstrated how the degree of individual contact between the vespertilionid *Myotis yumanensis* was a function of the temperatures in the
roost; under hotter conditions the bats were spread out, but when it
was colder they formed tight clusters. There are similar data for other
bats, usually vespertilionids such as *Antrozous pallidus* (Trune and Slobodchikoff 1976; Vaughan and O'Shea 1976).

In some cases bats may locate roosting aggregations by listening to
the echolocation calls of conspecifics (Barclay 1982a); other species,
such as *A. pallidus,* use directive calls (Brown 1976; Vaughan and
O'Shea 1976) which permit others to congregate in suitable roost sites.
As they enter roosts before dawn, *A. pallidus* spend considerable time
inspecting the location, so that in addition to the conspicuous directive
calls, the crowd of bats milling around the roost entrance provides a
wealth of visual and acoustic cues. The level of social structure within
the roosts, however, varies. Changes in position by one individual, or
arrivals of other bats, usually set off a series of vocal protests accompanied by visual displays such as baring of the teeth (Barclay, Fenton, and
Thomas 1979; Brown 1976). D. J. Howell (1979) noted that roosting
groups of the phyllostomid *Leptonycteris sanborni* were remarkably quiet
compared to vespertilionids, coinciding with evidence that at least

some of the bats roosting together also foraged together. Extended observation of individually marked *Myotis lucifugus* within a nursery roost (Thomson 1980) revealed no clear trends of preferred roost-mates or any evidence of a strong underlying social structure.

In aggregations of roosting bats, particularly those occurring in dark roosts, vocalizations are a common mode of communication. Variability in the social calls of bats may encode a great deal of information about the sender (Brown 1976; Fenton 1977). Unstructured aggregations of bats appear to be typical of (but not restricted to) species that enter torpor, particularly vespertilionids.

The richest communication repertoires among roosting bats occur in species with strongly structured roosting populations. A number of species have been studied in some detail, and they use a range of signals in communication.

McWilliam (1982) studied the roosting behavior of *Taphozous hildegardae* (Emballonuridae) in some caves in Kenya. The bats roosted in the twilight zones of the caves, thus making observations of their behavior relatively easy. Adult males did not form clusters, but maintained individual distances just beyond the striking range of an extended forearm. In this situation, if approached by another bat, the male stretched its head and body toward the intruder and at this time appeared to be sniffing. If the intruder was a female, the male approached her and sniffed her anogenital area. When the intruder was a male, the resident male became more aggressive, flicking its wing at the intruder by rapidly extending the forearm away from the side of the body, producing a rapid vibration of the folded end of the phalanx and the long third digit. Wing flicks were almost always directed at other males.

According to McWilliam (1982), a higher intensity of aggression involved attacks, with the territorial male leaning forward, extending the forearm and thumb, and striking toward the intruder. Territorial areas within the roost were scent-marked, usually by pressing the anogenital area to the substrate. Males often sniffed their territories, urinated on the substrate, and occasionally marked the areas with exudate from their throat glands. Males also rubbed the throat glands over the backs of females living in their territories, and marked themselves by rubbing their forearms and folded wings along the throat patch.

Territorial males with large groups of females and young herded them into tight clusters. The scent-marking and wing flicks were often accompanied by vocalizations audible to the human ear. McWilliam (1982) identified a "tcheek tcheek" vocalization produced by bats in warning postures with open mouth and head thrown back. The bats usually produced this vocalization during territorial squabbles, but also at the approach of danger. Another vocalization audible to McWilliam (1982) was a glissading series of squeaky notes, produced in association with no particular behavioral context.

The repertoire of *T. hildegardae* involves vocalizations, scent marks, and visual displays, including wing flicks and changes in posture. Similar repertoires have been found for other bats. Bradbury and Emmons (1974) studied another emballonurid in Trinidad. *Saccopteryx bilineata* males are territorial at roosts and within their territories; some males maintained harems. Individual males, with or without harems, are spaced at regular intervals in suitable roost sites, often between the buttresses of tree trunks in situations with good ambient light. Altercations between neighboring males involve mutual approaches to territory boundaries accompanied by short, high-frequency barks which may be simultaneously produced or exchanged. The barks are similar to echolocation pulses, but longer, and involve a jerk of the head. This

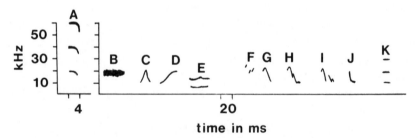

FIG. 22. Tracings of the sonagrams of the echolocation call (A) and some other vocalizations of the emballonurid *Saccopteryx bilineata*. Included are a bark (B), an inverted V call (C) typical of male hover displays which elicit sigmoidal (D) or harmonic series (E) calls from females, and the syllables found in the long and short songs of males, from pulses (F), hook notes (G), notched chirps (H), double chirps (I), L-notes (J), and whines (K). These are modified from Bradbury and Emmons (1974), who illustrated in more detail the sequences of syllables in long and short songs. Note the difference in time scale between the echolocation call and the other vocalizations.

type of interaction usually leads to a mutual withdrawal or to parallel movements along the boundaries (Fig. 22).

Bradbury and Emmons (1974) found that adult males used scent-marking in territorial situations. An adult male approached another bat, extended the folded wing on the side closest to the other bat, opened the orifice of the wing gland, and gave a series of short shakes of the wing while leaning out. Bradbury and Emmons called this pattern of behavior "salting." Males salted females, their own or from other harems, and other males along territorial boundaries. There was no evidence of marking territories with the anogenital area, but otherwise the patterns of behavior, corrected for different locations of glands (throat versus wings), are generally similar to those reported by McWilliam (1982); in both *Taphozous hildegardae* and *Saccopteryx bilineata*, spacing in roosts is maintained by multimedia displays.

According to Bradbury and Emmons (1974) there is a great deal of behavioral exchange as the bats return to their roosts at dawn; males return first, females later. As females approach, the males make audible (to humans) long songs, five to ten minutes at a stretch. The songs involve long series of cheeplike syllables. As the females land and occupy areas, males switch from the long songs to other displays. At this time salting is common. Males also perform hover displays in front of females, with their noses almost touching. During the hover display both males and females vocalize and the male may snap open his scent glands. These interactions appear to be greeting ceremonies.

Bradbury and Emmons (1974) observed no overt responses by females to salting behavior of the males. Once the bats settled into a roost situation, there was a period of inactivity and grooming, and then the males and/or females produced short songs at one- to five-minute intervals. Short songs involved five to thirty syllables, the same as the elements in the longer songs. Bradbury and Emmons found no evidence of counter-signing and noted that short songs might occur at any time of the day.

According to Bradbury and Emmons (1974), the long and short songs were produced by males with or without harems, and the long songs served an attraction function. Many of the displays were specific to males; females showed no bark, hover, or salting at boundary line displays. Females produced a range of vocalizations that elicited responses from males, as demonstrated by playback presentations of the female vocalizations. Females were aggressive to males and other females, and a female moving to another harem was usually attacked by the females already resident there. The attacks of females were made with folded wings in a manner similar to the attacks of males, but they were occasionally accompanied by nipping at the victim. Interactions between females were obviously complex and had a strong influence on harem composition. Observation of marked individuals revealed that some females were more prone to attack than others.

In *Saccopteryx bilineata*, the behavior of the males, particularly the displays they made at the roosts, influenced the composition of the roosting populations (Bradbury and Emmons 1974). In *Pipistrellus nanus*, an African vespertilionid, O'Shea (1980) found that adult males did not roost together, and that during the May to August dry season in Kenya, males called from their roost sites. Certain males were preferred by more females than were others, and there was considerable variation among males in their overall attractiveness to females. Harem size was highest in the most vocal males (O'Shea 1980). Again, the behavior of males and their displays affect the composition of roosting aggregations.

Hipposideros commersoni and *H. gigas* in Kenya also show strong structure in roost populations. In caves there, some males are territorial while others roost in bachelor groupings (McWilliam 1982). Adult males occupying territories defend areas on the ceiling or wall of the caves. Males lean forward to investigate an approaching bat, and in this posture show bursts of ear movements and obvious sniffing. If another male approaches too closely, the territorial male strikes out with the claw of the thumb, sometimes making contact, and occasionally pressing home his attack by biting with his canines. Males also scent-mark their territories by pressing the anal region repeatedly against the substrate or by rubbing it along ridges. Scent-marking was usually alternated with smelling of the treated areas. Other patterns of display in the roosts were associated with mating.

Porter has made other observations of social interactions among captive phyllostomids, *Carollia perspicillata*, and confirmed many of the patterns of behavior she observed by studies in the field (Porter 1978; 1979a; 1979b). Adult males with harems used a combination of vocalizations and wing displays, which Porter called "boxing," to keep other males away from their harems. Males also used approaches, wing pokes, and vocalizations to keep members of their harems together in space, and harassed them via these displays until they formed clusters. Other patterns of behavior involved displays to females associated with mating.

When another male approached the space occupied by a harem male, the resident male approached the boundary and there began to shake his wings and vocalize at the intruder. Male-male confrontations tended to be stereotyped, involving alternate strikes with wings, during which the wings and mouth were partly open. This behavior was accompanied by loud, harsh vocalizations. These displays were usually performed by males with enlarged testes.

Porter found little evidence of female-female interactions that might have strongly influenced roosting associations, results contrasting sharply with those for *Saccopteryx bilineata* (Bradbury and Emmons 1974). Other harem-forming bats, such as *Phyllostomus hastatus,* show high levels of aggression between males holding harems. The aggressive displays include vocalizations, wing-beating, and fights which involve biting (McCracken and Bradbury 1981); such displays are typical of males and have not been reported among females. In other species presumed to have harems, for example the phyllostomids *Vampyrum spectrum* (Vehrencamp, Stiles, and Bradbury 1977) or *Trachops cirrhosus* (Tuttle 1982), only one group may occupy a roost site, presumably limiting aggressive interactions between competing males.

The influence of aggression between males has strong implications for the success of harem males. McCracken and Bradbury (1977) showed that harem males in *Phyllostomus hastatus* sire most of the offspring born in their harem, and Porter's (1978; 1979a; 1979b) observations of captive *Carollia perspicillata* indicate a strong involvement of the male with the offspring born in his harem.

Within their day roosts, then, there is a range of communicative interactions among bats, usually involving multimedia displays which may strongly influence the distribution of females and males in the roosts. In some cases, vocal or visual displays such as baring of the teeth are associated with the jostling of individuals as they arrive, depart, or change their positions in a roosting assemblage. Some patterns of communication are probably instrumental in mutual avoidance, while other signals and displays lead to aggregation. The spectrum of conditions is continuous, reflecting variation in the social nature of roosting bats and differences in roost conditions, from fully lighted to totally dark.

MATING BEHAVIOR

There have been relatively few detailed studies of the mating behavior of bats, but the available data suggest a two-tiered communication system in some species. At one level are displays that serve to attract potential mates, and at the other level are signals that ensure appropriate mood and specific identity of partners. Bradbury (1977a) reviewed the available data and pointed out that species living in dispersed situation make more use of the first level of signals than do gregarious forms. The situation involving gregarious species is complex, reflecting the various social conditions encountered among roosting bats. As Bradbury (1977a) pointed out, several mating systems are known from the Chiroptera. Evidence for monogamy appears to be relatively circumstantial, and many species are polygynous, with males forming and defending harems. Gradations toward a lek system are known among epomophorine bats, with the lek occurring in *Hypsignathus monstrosus* (Bradbury 1977b; 1981). In at least one species, the vespertilionid *Myotis lucifugus,* there is no evidence of a structured mating system, and mating appears to be random and promiscuous (Thomas, Fenton, and Barclay 1979).

The best examples of displays serving to advertise the location and availability of individuals are known from male epomophorine bats. Adult males in these African species, both dispersed individuals and those aggregated in communal display areas, typically use vocal signals (Bradbury 1977b). The calls of males of several epomophorines are known (e.g., Rosevear 1965; Kingdon 1974; Wickler and Seibt 1976), making them conspicuous to people throughout much of their range (Fig. 23). The available evidence suggests that this calling behavior, at least in the range of frequencies audible to man, is not common or has not been recognized among other bats, although the Old World vespertilionid *Nyctalus noctula* is an exception (Bradbury 1977a).

The most complete data about calling behavior of male epomophorines comes from the work of Bradbury (1977b) on the lek-forming *Hypsignathus monstrosus.* In a lek, females come to display areas where males aggregate and there select a male with whom they copulate. The only resource at the site is the male, and some males are much more sexually active than others. There is also striking sexual dimorphism and no paternal involvement in the raising of young. The mating system of *H. monstrosus* meets all of these criteria.

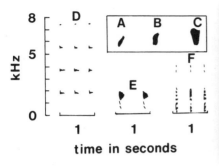
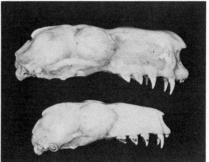

FIG. 23. A comparison of the relative sizes of the larynges in three epomophorine bats, *Epomophorus wahlbergi* (A), *Epomops franqueti* (B), and *Hypsignathus monstrosus* (C), and the calls used by males of these species to attract females (D, E, and F, respectively). Also shown are the skulls (G) of male (top) and female (bottom) *H. monstrosus* to illustrate the sexual dimorphism in this species. Data on larynges and vocalizations modified from Wickler and Seibt (1976); photographs reproduced with permission of the University of Toronto Press.

During the breeding season male *H. monstrosus* aggregate in linear assemblages, usually along the banks of rivers, where they call from roosts in trees. The vocal display of the males consists of a monotonous (to us) series of honks which continue through the night and which are accompanied by inflation of the vocal sacs and flapping of the wings. The rate of production of vocalizations increases as females pass by, and this change in behavior permitted Bradbury (1977b) to assess the numbers of males involved in the lek and the incidence of females passing through the areas defended by individual males. Furthermore, a characteristic release call (produced by the females) that followed mating allowed measurement of mating success rates of different males involved in the lek. The vocalizations of the males were clearly important in attracting females, but there is no evidence about the role of other components of the display in the choices made by females.

Wickler and Seibt (1976) reported other observations on calling of epomophorines, specifically *Epomophorus wahlbergi*. This species is characterized by more-dispersed males that call from predictable sites and increase their rates of calling in response to fly-bys by females. Wickler and Seibt did not find evidence of a release call in this species, but it appears that the vocal displays of males serve to advertise them to females. Bradbury (1981) found that another species of epomophorine bat, *Epomops franqueti*, shows looser aggregations of calling males, with greater distances between males than those reported for *H. monstrosus*. These "exploded leks" correlate with differences in the diets of females and with the size of the bats (Bradbury 1981), and again males increase their calling rates in response to visits by females. In *Epomops buettikoferi* the males are also more dispersed in calling sites and also increase their rates of display when females fly past (D. W. Thomas, pers. comm.).

Observations of the vespertilionid *Nyctalus noctula* suggest that males occupy hollows in trees, calling from these sites during the mating season, presumably to attract passing females to their harems (Likhachev 1961; Sluiter and van Heerdt 1966). Detailed studies of marked individuals are lacking, however, and more information is required to clarify the situation. Morrison (1979) found that male *Artibeus jamaicensis* (Phyllostomidae) defend hollows in trees, presumably to gain access to females roosting there. Kunz, August, and Burnett (1983) found evidence of harems in *A. jamaicensis* roosting in caves in Puerto Rico.

Our knowledge of bat displays that serve to attract members of the opposite sex is very limited. Most of the data are based on observations made possible by displays which are conspicuous to human observers, although Morrison's (1979) data arise from telemetry studies. Since lower-frequency sounds carry better than higher-frequency ones, one might expect that bats using acoustic displays to advertise their presence would resort to lower-frequency calls. The earlier comments about the role of echolocation calls in communication, however, may be highly relevant in mate-seeking bats.

Although there are some detailed observations of mating and the behavior associated with it, the data base is limited. Most of the detailed observations and descriptions suggest that during and before copulation, signals transmitted by sounds, scents, and tactile stimuli are important.

Nelson (1965) reported in detail the mating behavior of the pteropodid *Pteropus poliocephalus*. Adult males established territories around females in day roosts (camps), often including one or more females in their defended space. As noted above, the territory was marked with the male's scent and was defended by the male's use of a mélange of behavior patterns, including vocalizations and physical attacks on intruders. Females participated less in territory defense than did males, but they would initiate a defense and leave the matter to be settled by the resident male.

The male approached females in their roosts and, when his approaches were tolerated, came as close as possible, trying to lick the genital area of the female. Continued tolerance of his presence led to mutual grooming, which, with licking of the female's vulva, intensified as the mating season progressed. Copulation was usually initiated by the male, but sometimes by the female. To copulate, the male approached the female from behind and enfolded her with his wings, a posture common among the Chiroptera. During and immediately before copulation, the male showed a "frightened" facial expression with his ears laid back.

Neuweiler's (1969) observations of mating in *Pteropus giganteus* support this general pattern of mating behavior, but he also found a typical and distinctive copulation call associated with mating. Unlike the seasonal colonies of *P. poliocephalus,* those of *P. giganteus* were permanent, with individuals occupying the same roosts permanently. Males appeared to be more committed to particular roosts than did females, who often moved between roosts from day to day. In *P. giganteus,* the vertical position in the roost indicated the position of the individual in the dominance hierarchy.

Where males are associated with the same females through the mating season, precopulatory displays probably serve to facilitate individual recognition, and the data from the two species of *Pteropus* provide examples of this. In contrast, mating in *H. monstrosus* involves fleeting contact between males and females, typical of lek situations, and there is no conspicuous prelude to copulation once a female has made her choice and presented herself to a displaying male (Bradbury 1977b).

Male *P. poliocephalus* often scent-mark the female(s) in their roost territory, a pattern of behavior described from some other bats, but they do not necessarily do this to denote "ownership." McWilliam (1982) described mating behavior in *Hipposideros commersoni* and *H. gigas,* based on his observations of interactions at territories in roost sites defended by adult males. A male approached a female that entered his territory and directed rapid eversions of the frontal sac gland toward her. If she remained in the area, he approached more closely and nosed her genital area, alternating his inspection with eversion of the frontal sac gland. Males performed wing-flap displays interspersed with inspection behavior and eversions of the frontal sac gland to receptive females. In copulation, the male mounted the female from the rear while continuously flapping his wings. After mounting, the male leaned forward, everting his frontal sac gland over the female's nose leaf, and vocalizing. At this point, male *H. gigas* squeezed the female. During copulation in both species, the male crossed his fore-arms to pin the female's arms to her side, and lifted her from the surface. At this stage the males ceased to evert the frontal sac gland, but continued to vocalize. The entire mating sequence usually took less than 20 minutes, and afterward both sexes groomed their genital areas. McWilliam (1982) found no evidence of interference in mating by neighboring males, and proposed a mating system operating on female choice. Harems averaged two females; the largest observed was five. Females not immobilized by males before copulation could stop the attempted copulation by struggling or foil the male by folding the interfemoral membrane forward.

   McWilliam's (1982) description of mating emphasizes the impor-tance of tactile, olfactory, and auditory components in the mating interactions of *H. commersoni* and *H. gigas*. He compared eversion of the frontal sac gland to eversion of glands by other bats, including the epaulettes of male epomophorine bats (e.g., Wickler and Seibt 1976), or the antebrachial glands of male *Saccopteryx bilineata* (Bradbury and Emmons 1974). He also pointed out that the other behaviors, including nuzzling of the genital region of the female and scent marking, as well as pinioning of the female, were recurring patterns of mating behavior among the Chiroptera. In a situation where males may consort with different females, or females with different males, the kinds of be-havior patterns described by McWilliam (1982) could serve to ensure appropriate conditions of mood and receptivity rather than individual recognition.

   Porter (1979a) described mating behavior in the phyllostomid *Carol-lia perspicillata*, usually involving females mating with the male in whose harem they roosted during the day. A male with enlarged testes and partly opened wings approached a female, extending his head toward her, apparently trying to sniff her. The male then prodded the female with an extended wing, and if she remained passive, would lick her or bite her on the back of the neck before wrapping his wings around her and copulating from the rear. Before any close approach the male rapidly flicked his tongue in and out and intermittently produced loud, harsh vocalizations. Before mating, males sometimes hovered in front of females and vocalized at them. The pattern is again one of multimedia interactions. The tongue flicking is reminiscent of that described by D. J. Howell (1979) for *Leptonycteris sanborni* in another social context, and suggests involvement of the Jacobson's organ.

The mating of *Myotis lucifugus* involves fewer precopulatory interactions between male and female. Mating may occur in two phases, an active one in the latter part of the summer, and a passive one during the winter when most of the bats are hibernating. During the active phase, males and females are alert and active, but during the passive phase active males seek out and mate with torpid females. This pattern of mating appears to apply to other hibernating vespertilionids (e.g., *Plecotus townsendii;* Pearson, Koford, and Pearson 1952), and perhaps to rhinolophids (e.g., Ransome 1980). Thomas, Fenton, and Barclay (1979) described the behavior associated with mating in *M. lucifugus*. During the active phase, males occupy crevices or holes high on the walls or on the ceilings of hibernacula, usually caves or mines, and from there emit vocalizations typical of echolocation. Females fly up and down the passages of the hibernacula, stopping to stay with males. Clusters of active bats accumulate around adult males, as indicated by captured groups at mating sites. Individual groups typically include one adult male with several adult females and some subadults of both sexes. In *M. lucifugus* females may be sexually mature in their first autumn, while males are not (Herd and Fenton 1983).

A bat arriving at one of these groups is met by the adult male in naso-nasal greeting. When the new arrival is another male, a brief scuffle may occur before the newcomer leaves. Naso-nasal greetings would expose the secretions of the enlarged pararhinal glands to both participants. Thomas, Fenton, and Barclay (1979) found no evidence of precopulatory displays other than the naso-nasal contacts, and interaction between males and females was brief, involving only the time associated with copulation, up to 20 or 30 minutes.

During the passive phase, males moved through the hibernacula investigating clusters of torpid bats and occasionally attempting to copulate with one of the torpid animals. The accosted bat might or might not arouse from torpor, and in some cases was a male. While copulating with active and struggling females, males sometimes produced a distinctive copulation call (Barclay and Thomas 1979). During

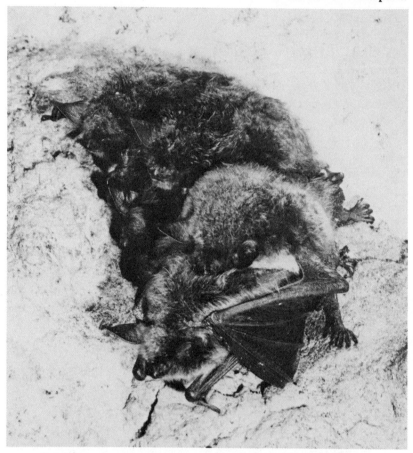

FIG. 24. A male *Myotis lucifugus* copulating with a torpid female
while other torpid individuals remain in hibernation in the background. The male is using his teeth and thumbs to hold on to the
female. Photograph reproduced with Permission of the University
of Toronto Press.

mating the male approached the female from behind, typically em-
braced her with his forearms, and grabbed her by his teeth at the scruff
of her neck (Fig. 24). The incidence of copulation calls increased in
direct proportion to the amount of resistance the male met in his
mating attempt.

The observations of Pearson, Koford and Pearson (1952) of mating
in the vespertilionid *Plecotus townsendii* suggested that females might
remain torpid through the entire copulation. They reported no

homosexual matings and found that active bats in captivity included precopulatory displays in their repertoires. The mating display they described involved the male hanging in front of the female and producing a twittering vocalization, as well as extensive scent-marking of the female with exudate from the facial glands. The male marked the female's neck, face, forearms, and ventral surface. Unfortunately, Pearson, Koford, and Pearson (1952) provided relatively few observations of actual matings and no indication of the nature of the mating system of these bats.

There is a trend within the Chiroptera for more elaborate mating systems to be associated with more sophisticated communication behavior. The mating system of *M. lucifugus* appears random and promiscuous (Thomas, Fenton, and Barclay 1979), apparently because neither sex can protect any investment made in mate choice. There is no evidence of elaborate prenuptial displays beyond naso-nasal contact which might serve to identify the sex of the other bat. Attempted homosexual matings suggest lack of olfactory information and mistaken identity. In species with more structured mating systems precopulatory behavior is more diverse and involves a range of stimuli serving in individual recognition or to identify the moods of the animals involved. Lack of data, however, makes wide generalizations premature.

MOTHER-YOUNG INTERACTIONS

Although some bats are commonly cited as examples of mammals in which mothers do not selectively nurse their own young (e.g., Wilson 1975), most species studied to date recognize and nurse their own offspring. The proposal that mother bats nurse indiscriminately is based on weak evidence from the molossid *Tadarida brasiliensis* (Davis, Herreid, and Short 1962) and the vespertilionids *Miniopterus australis* and *M. schreibersi* (Brosset 1962); in this case "weak" evidence reflects the lack of detailed information about individually recognizable mothers and offspring. There are now data on mother-young interactions in a range of species of bats, and in most situations females recognize and selectively nurse their own young (Table 7); but females also occasionally nurse young other than their own. The unanswered question is whether selective nursing is the rule or the exception.

Not all studies are in agreement, however, and Roth (1957) switched

TABLE 7. Summary of studies of mother-young interactions in bats

FAMILY / SPECIES	NATURE OF STUDY	MODE OF RECOGNITION	NURSED OWN YOUNG?	SCENT-MARKED YOUNG?	SOURCE
Pteropodidae					
Pteropus poliocephalus	field observation and experiments	vocalizations and olfaction	yes	no	Nelson 1964; 1965
Pteropus giganteus	field observation	vocalizations and olfaction	yes	no	Neuweiler 1969
Rousettus aegyptiacus	lab observations and experiments	vocalizations and olfaction	yes	no	Kulzer 1958; 1961
Rhinolophidae					
Rhinolophus ferrumequinum	lab observations	vocalizations	yes	no	Möhres 1967
	lab and field	vocalizations	yes	no	Matsumura 1979; 1981
Megadermatidae					
Megaderma lyra	lab	isolated young called	?	?	Novick 1958
Phyllostomidae					
Macrotus californicus	lab observations	vocalizations and olfaction	yes[1]	no	Gould 1977
Phyllostomus hastatus	lab observations	vocalizations and olfaction	yes	no	Gould 1977
	field		yes	no	McCracken & Bradbury 1977
Leptonycteris sanborni	lab observations	vocalizations	yes	no	Gould 1977

Species	Method		ColA	ColB	Reference
		tion			
Desmodus rotundus	lab		yes	no	Kleiman & Davis 1979
	lab observations	vocalizations and olfaction	yes	no	Schmidt 1972; Gould 1977
Vespertilionidae					
Myotis lucifugus	field observation and playback	vocalizations and olfaction	yes	no	Thomson 1980
	lab observation and experiment	vocalizations and olfaction	yes	no	Gould 1971; Turner, Shaughnessy, & Gould 1972
Myotis velifer	field observation	vocalization and olfaction	yes	no	Kunz 1973
Myotis daubentoni	field observation	vocalization and olfaction	yes	no	Swift 1981
Pipistrellus pipistrellus	field observation	vocalization and olfaction	yes	no	Swift 1981
Eptesicus serotinus	lab		yes	no	Kleiman 1969
	lab observation	vocalization and olfaction	yes	yes	Kleiman 1969
Nyctalus noctula	lab observation	vocalization and olfaction	yes	no	Kleiman 1969
Eptesicus fuscus	field observation	vocalization and olfaction	yes	?	W. H. Davis, Barbour, & Hassell 1968
Nycticeius humeralis	field observation of marked animals	vocalization and olfaction	yes[2]	yes	Watkins & Shump 1981
	lab observation		yes		
Plecotus townsendii	field observation	vocalization and olfaction	yes	?	Pearson, Koford, & Pearson 1952
Miniopterus schrebersi	field	young called when isolated	no	?	Brosset 1962

TABLE 7. (*continued*)

FAMILY SPECIES	NATURE OF STUDY	MODE OF RECOGNITION	NURSED OWN YOUNG?	SCENT-MARKED YOUNG?	SOURCE
Miniopterus australis	field	young called when isolated	no	?	Brosset 1962
Antrozous pallidus	lab experiments	vocalization and olfaction	yes	no	Brown 1976
Molossidae	field observations		yes	no	Davis 1969
Tadarida brasiliensis	field observations	vocalizations and olfaction	no	no	R. B. Davis, Herreid, & Short 1962
Tadarida condylura	field observations	vocalizations and olfaction	yes	?	Kulzer 1962

[1]one record of female taking young not her own; [2]own young nursed preferentially during early development

infant *Myotis lucifugus* between females held in separate cages and found that the infants continued to grow and remained healthy. He concluded that selective nursing did not necessarily occur in *M. lucifugus*. Subsequent work on this species in the laboratory (e.g., Gould 1971; Turner, Shaughnessy, and Gould 1972) and in the field (Thomson 1980) clearly indicate that mothers can and do recognize their own offspring.

The most recent research into the situation in *Tadarida brasiliensis* is a genetic analysis of pairs of mothers and young found together with the young nursing. In this way McCracken (1984) established that in 83 percent of the 167 mother-young pairs they examined, females nursed young that genetically could have been theirs; in 17 percent of the cases females nursed young that could not have been theirs (Fig. 25).

The situation is not completely clear. Watkins and Shump (1981) found that during the initial period of development female *Nycticeius humeralis* (Vespertilionidae) selectively nursed their own offspring, but

FIG. 25. This photograph of a crèche of young *Tadarida brasiliensis* emphasizes the problems faced by females of this species when searching for their own offspring. Photograph courtesy of Gary F. McCracken.

as the young became more active the females nursed any offspring that approached them. Furthermore, under some circumstances females assist and nurse offspring which are not their own: e.g., Gould's (1977a) example of this from the phyllostomid *Macrotus californicus* under captive conditions. In socially structured nurseries adult females other than the mother, and in some cases the harem males, attempted to influence females that had dropped their babies (Porter 1979a). The details of these situations will be considered below.

The situation facing mothers is not uniform among bats. Many solitary species leave their young in the day roost when they leave the roost to feed, and later return to their young. An example of this setting is provided by most lasiurine bats of the New World. Solitary fruit-eating bats may carry small babies with them when they feed, eliminating, at least initially, the need for detailed mechanisms of recognition. Some species which occupy day roosts with conspecifics move infants to alternate roosts before departing for the evening's feeding, e.g., *Saccopteryx bilineata* (Bradbury and Emmons 1974). In other situations where the young is left behind in a nursery, it may be left alone. In some colonial species small young may be carried, particularly among frugivores (e.g., Kleiman and Davis 1979); yet other colonial species do not carry their young when they go to feed (e.g., Nellis and Ehle 1977). Other species of bats deposit their young in crèches (e.g., *Myotis nigricans,* D. E. Wilson 1971; *Myotis thysanodes,* O'Farrell and Studier 1973; *Macrotus waterhousii,* Goodwin 1970), leaving them in the day roost in large clusters, usually in the care of an adult. It is not surprising that examples of apparent indiscriminate nursing involve species forming huge colonies where tens of thousands of babies are left when their mothers depart to feed (Fig. 25).

There are at least three categories of problems in mother-young interactions. When the infant has been left alone or with its siblings of the same age, the mother must find the roost where she left her young. If the mother moved the offspring to another roost away from the communal day roost, the problem is to remember where the alternate roost is located. If the young have been left alone in some location in the day roost, locating it may be easier than if it was placed in a cluster with many other babies. Clearly the magnitude of the problem depends upon the mobility of the young and the size of the colony. Where the offspring is left in a crèche with many others, the problem centers around identification of the correct infant(s) in the mass. In this situation, the scale of the problem depends upon the size of the colony or crèche and is influenced by the degree of coincidence of parturition, which affects the number of same-age infants in the colony.

Female bats could use several cues to find their young, and baby bats to recognize their mothers. Bats have well-developed senses of spatial memory (e.g., Bradbury 1977a; Mueller and Mueller 1979; Gaudet 1982), which would serve females well in locating the site where young had been deposited. Spatial memory could easily permit a female to come within a meter of the location where she had left her young (e.g., Gaudet 1982).

Having located the roost, females may use olfactory and/or auditory cues to identify their offspring. Differences in roosting habits could produce interesting variations in the behavior of young, which could be relatively easily tested under some circumstances. Females roosting alone may show less refined recognition of their own offspring, since the main problem is finding the roost site. Solitary young should benefit from being as inconspicuous as possible, and their tendency to call to attract their mothers may be much lower than that of colonial species where young may be left with adult protection. Unfortunately our knowledge of the behavior of baby bats of solitary species is limited, but R. M. R. Barclay (pers. comm.) found that infant *Lasiurus cinereus* left alone when their mother went to forage called almost continuously.

At this point it is appropriate to consider the details of a mother-young reunion, and the situation in *Myotis lucifugus* (Thomson 1980) appears to be typical. Thomson placed young on retrieval posts in the attic colony she studied, and observed the behavior of the adults and young via a total darkness television camera to minimize disturbance to the bats. Typically isolated infants were investigated by several passing females, some of whom merely flew close to the youngster, while others landed beside it, sniffed at it, and then either retrieved it or flew away. Playback experiments indicated that the amount of attention paid to a calling infant did not reflect the numbers of infants calling from a location. The detailed behavior of the infant depended partly upon its age. Very young bats tended to call continuously when left by their mothers, and they always responded with isolation calls to playback presentations of echolocation calls or double notes. There was a slight tendency for the double-note calls to attract more attention (elicit a greater response) from infants than echolocation calls. Older but still unweaned infants tended not to call continuously, and responded to only some passing females, or playback presentations. Thomson's (1980) data indicate that the interaction between mother and offspring changes, that older offspring apparently recognize their mothers, and that reciprocal identification is clearly involved. A very important point, however, is that while young attempted to nurse from any female, mothers were very reluctant to feed strange young, a pattern typical of many mammals (e.g., Spencer-Booth 1970). Thomson's (1980) data underscore the importance of olfaction and vocalizations in mother-young recognition, and appear to be representative of many species studied to date (Table 7).

There are two obvious sources of olfactory information for the mother-young reunion, one involving odors from the young acquired passively through the milk or saliva of the mother, the other involving specific scent marks. From birth baby bats are commonly licked by their mothers, often in grooming, and females could easily impart to their offspring diagnostic odors via saliva or milk. Recognition mediated by specific odors is involved in many species of bats, and Kulzer (1958) showed that mother *Rousettus aegyptiacus* could recognize their own young by olfactory cues alone. A few species show evidence of scent-marking of their young.

Watkins and Shump (1981) observed that female *Nycticeius humeralis* marked their offsprings' faces with exudate from submaxillary glands before evening departure. Within three days of parturition this gland increases dramatically in size and it regresses prior to weaning. Regression of the gland coincides with the time when females start to nurse any young which approaches them. The most detailed information about this phenomenon is provided by Watkins and Shump (1981), but Kleiman (1969) found a similar situation of glandular activity associated with lactation in the vespertilionid *Eptesicus serotinus*. The less detailed observations of Stebbings (1966) of glandular activity in *Plecotus* spp. (Vespertilionidae) could indicate marking of young by glandular products, but the timing is not right. In the North American *Plecotus townsendii* secretions from facial glands are associated with mating (Pearson, Koford, and Pearson 1952). Species for which there is evidence of scent-marking of young with glandular exudates are shown in Table 7. Olfactory cues are important in proximal recognition of young by their mothers, but only in some cases are the cues associated with specific scent-marking by glandular products.

Vocalizations also play a vital role in mother-young recognition, and virtually all of the species studied to date have specific vocalizations for this in their repertoires (Table 7). The importance of vocalizations in mother-young interactions is demonstrated by the work of Gould (1971; 1977a), who found that female *Eptesicus fuscus* ignored silent or abnormally vocalizing young. Vocalizations were clearly essential in eliciting proper maternal behavior in this species, and the same is true in *Noctilio albiventris* (Brown, Brown, and Grinnell 1983). Young commonly produce isolation calls, and both young and their mothers may produce double-note calls as well (e.g., Gould 1977a).

Gould (1977a) has proposed the following model of mother-young reunions: 1) mothers leave the roosts, and infants emit double-note calls at low rates of repetition (this is typical of phyllostomids, vespertilionids usually produce isolation calls); 2) mothers return to the roosts and emit FM pulses as they fly, infants respond with double-note or isolation calls to alert their mothers to their presence; the infants may also produce FM calls; 3) mothers approach infants, and infants respond by increasing their rates of calling; the FM pulses facilitate location of the infant by the mother; 4) each mother determines by olfactory and/or acoustic cues the identity of her infant(s) and guides it (them) to her nipple. This model is compatible with the results of Brown (1976) for *Antrozous pallidus* and Thomson (1980) for *Myotis lucifugus*, although both of these vespertilionids rely on isolation calls more than on double notes.

This model emphasizes the importance of FM calls usually associated with echolocation by adult bats. Gould (1977a) pointed out that these FM calls are easily accurately located. If the data of Matsumura (1979; 1981) are representative, female *Rhinolophus ferrumequinum* use only echolocation calls as they approach their infants, indicating that specialized vocalizations such as double notes are not always involved in mother-young reunions. There are similar observations for the African molossid *Tadarida condylura* (Kulzer 1962). Matsumura's data, however, show that reunions between mothers and offspring in *R. ferrumequinum* progress from asynchronous to antiphonal to simultaneous calling, with both mother and offspring producing vocalizations typical of echolocation. These observations emphasize the importance of echolocation calls in communication roles, and Thomson's (1980) playback experiments suggest that baby bats may recognize the echolocation calls of their mothers.

Different patterns of isolation calls of young of different species

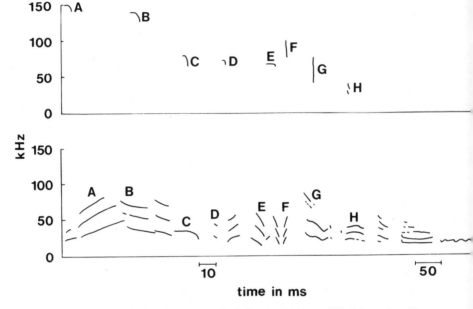

FIG. 26. Vocalizations of some Malaysian bats modified from Gould
(1980). Included here are the echolocation (above) and infant
isolation calls (below) of *Hipposideros cineraceus* (A), *H. bicolor* (B), *H.
ridleyi* (C), *H. armiger* (D), *H. diadema* (E), *Nycteris javanica* (F),
Megaderma spp. (G), and *Taphozous* spp. (H). Note the 50 ms time
scale only applies to the infant isolation calls of H. In the case of
Hipposideros spp., infant isolation calls of species roosting together
are strikingly different, presumably to minimize confusion during
mother-young reunions.

could facilitate mother-young reunions in settings where several species share a roost. Gould (1980) has documented this kind of interspecific variation in a nursery roost in Malaysia (Fig. 26). There is also
considerable variability in the isolation (Fig. 27) and double-note calls
of individuals, and some observers have used this variability to predict
that these vocalizations contain vocal signatures (e.g., Brown 1976;
Gould 1977a; Thomson 1980). In some cases the results of playback
experiments indicate considerable levels of accuracy in the mothers'
identification of their young by the youngs' vocalizations (Brown 1976;
Turner, Shaughnessy, and Gould, 1972; Thomson 1980). In the non-
echolocating megachiropterans, vocalizations also play a vital role in

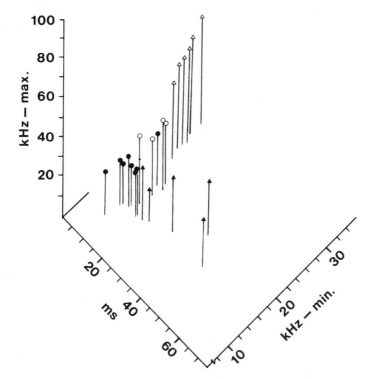

FIG. 27. Compared here are the major power band (kHz-max.), the minimum frequencies (kHz-min.) and the durations (ms) of the isolation calls of four individual *Myotis lucifugus* (Δ 17 mm forearm; ● 18 mm forearm; ○ 22 mm forearm; and ▲ 23 mm forearm). The comparison shows how differences in these features could combine to permit individual recognition by vocalizations (modified from Thomson 1980).

mediating mother-young reunions (e.g., Kulzer 1961; Nelson 1964; Neuweiler 1969).

The accumulated evidence about the recognition of young by their mothers in bats makes clarification of the situations where indiscriminate nursing has been proposed essential to the broader picture. There are occasional records of mother bats taking young that were definitely not their own (Gould 1977a), albeit in a situation where there was no information about the possible degree of relatedness between participating bats. Against this background, the observations of Porter (1979a) in a captive colony of *Carollia perspicillata* are relevant. She found that a harem male and other female members of the harem scolded and wing-poked a female not responding to the calls of her isolated infant, a behavior Porter could elicit by playback presenta- tions. In one case, a male from an adjoining harem, the one previously occupied by the female in question, scolded and harassed the female until she retrieved her young. Brown (1976) reported that *Antrozous pallidus* in a captive social unit produced directive calls aimed at a mother responding to the isolation calls of a baby that was not hers.

These observations for two species of bats underscore the degree of social structure existing in some roosts and make it clear that vocaliza- tions play an important role in interactions between mothers and their young. These observations are unique to date, and Thomson (1980) found no evidence of the behavior reported in detail by Porter (1979a) or suggested by Brown (1976) in her detailed study of *Myotis lucifugus.* Bats with looser social organization in nursery colonies, e.g., *M. lucifu- gus,* may not show this behavior.

In summary, recognition of young bats by their mothers is the rule rather than the exception, and the evidence for milk herds is not strong. Recognition appears to be a reciprocal operation (e.g., Kulzer 1958; Nelson 1965; Thomson 1980), but in many cases it may turn out to be the females that show a high level of selectivity. The mechanisms for recognition include olfactory and auditory cues, and in some cases the olfactory cues are provided by specific scent-marks. Auditory signals of adults and young include echolocation calls, and some species supple- ment these with double-note or isolation calls. The role of spatial memory in mother-young retrievals warrants further study, and the differences in roosting patterns may correlate with important differ- ences in mother-young interactions.

VOCALIZATIONS

Relatively complete vocalization repertoires are available for very few species of bats. Studies are available for *Pteropus poliocephalus* (Nelson 1964), *Pteropus giganteus* (Neuweiler 1969), *Myotis lucifugus* (Barclay, Fenton, and Thomas 1979), and *Carollia perspicillata* (Porter 1979a; 1979b) (Fig. 28). Other papers deal with some of the vocalizations of some bats (e.g., Bradbury and Emmons 1974; Brown 1976; Gould 1980; O'Shea 1980).

There is considerable variety in the calls of bats, generally within the predictions of Morton (1977), with harsh calls associated with aggres-

FIG. 28

	sonogram	average duration in ms	bandwidth in kHz	major power band in kHz	harmonics
a-*Myotis lucifugus*					
echolocation call		4	80–40	n/a	+/–
echolocation call + honk		4	80–30	n/a	–
short squawk		62±61	<1–30	28–17	+
long squawk		591±238	<1–35	29–18	+
audible buzz		1–3	<1–100	28–18	+/–
squeak		28±27	15–40	n/a	+
double note		30±7	40–120	n/a	+
short isolation call		5±1.3	20–8	n/a	+
long isolation call		21±9	16–50	n/a	+
copulation call		61±14	7–17	n/a	+
sine wave call		>500	30–50	n/a	–
long squeak		50–100	25–80	n/a	+/–

Fig. 28 (continued)

b-*Carollia perspicillata*	sonogram	average duration in ms	bandwidth in kHz	major power band in kHz	harmonics
echolocation call		1	16–64	n/a	+/–
whine		140–500	8–30	n/a	+
warble		140–300	5–64	n/a	+
trill		60–180	8–32	n/a	–
screech		140–400	16–32	n/a	–
double note		30–40	8–35	n/a	–
FM glide		153	8–64	n/a	+
back checks		7–17	16–50	n/a	–
back checks grade to trills		42–63	20–50	n/a	–
buzz	sequence of FM pulses ending with FM glide				

Fig. 28. The vocal repertoires of three bats, *Myotis lucifugus*, *Carollia perspicillata*, and *Pteropus poliocephalus*, giving an impression of the range of vocalizations encountered in these species. The purpose of this presentation is to provide an indication of the appearance (sonogram) of the vocalizations, their durations and frequency ranges, rather than specific details about fine structure. Modified from Barclay, Fenton, and Thomas (1979) and Thomson (1980) for *M. lucifugus*, Porter (1979b) and Gould (1977) for *C. perspicillata*, and Nelson (1965) for *P. poliocephalus*.

c-Pteropus poliocephalus

isolation call		100–800	2–7	n/a	+
contact		50	6–8	n/a	
isolation trill		120	1–14	n/a	+
location		30–70	2–14	n/a	+
threat		350–500	2–14	n/a	–
female searching		200–400	1–14	n/a	+
female contact		50–150	1–10	n/a	+
short call		50	2–9	n/a	
long call		180–350	1–14	n/a	
threat a		10–30	5–9	n/a	
threat b		50–70	1–14	n/a	
female wing-flap		200–500	1–14	n/a	
male wing-flap		450–650	1–14	n/a	
precopulation call		10–50	1–14	n/a	
copulation call		300–400	1–14	n/a	
territory		300–500	2–14	n/a	
alarm		100–300	2–12	n/a	
feeding		250	2–14	n/a	

aerial predator—not recorded, but short, loud, harsh cry

avoidance—not recorded, but similar to aerial predator except for reaction from other bats

sive intentions and tonal calls with friendly or fearful ones. Several studies suggest considerable variation in the calls of individuals (e.g., Brown 1976; Fenton 1977; Barclay, Fenton, and Thomas 1979; Thomson 1980), and such findings have important implications for studies of communication. Variability in the vocalizations of individuals potentially allows the communication of a great deal of information concerning the sender and/or the situation.

The calls of bats can be arrayed into a variety of categories: echolocation, warning-aggression, sexual, and mother-young (Fig. 29). There are two basic patterns of echolocation calls: broadband clicks, and calls which show changes in frequency over time. There is a record of a clicklike vocalization being used by *Myotis lucifugus,* apparently for echolocation (Barclay, Fenton, and Thomas 1979), but details are lacking. It is obvious that echolocation calls can serve a function in communication.

Vocalizations used by bats in a warning or aggressive context tend to be broadband and harsh. These calls can be broadband and screechlike, with little in the way of tonal components, or composed of a series of short, tonal FM sweeps, producing a buzzlike effect, a recurring theme among many bats. Buzzlike vocalizations are often associated with hostile male-male interactions, for example between adjacent harem males (Porter 1979b), or at territories (Nelson 1964). The same basic call may serve in territorial defense, by an individual at a feeding location, or as an alarm call (Nelson 1964).

Sexual calls include a range of tonal vocalizations which may serve to attract females to males, as greetings between males and females of the same harem, or just before or during copulation. These calls tend to be tonal, sometimes rich in harmonics, and in some situations they are used in conjunction with other displays, including, for example, flight. Gould (1983) has pointed out that some of the vocalizations used in sexual encounters have precursors in form among calls used in mother-young interactions.

Tonal calls, often rich in harmonics, are also associated with mother-young interactions. Isolation calls produced by young separated from their mothers are common, although in some cases the young and their mothers use double-note calls in this situation. Particularly among vespertilionids the young produce isolation calls, and their mothers double-note calls. Within the Chiroptera for which there are data there is considerable variation in the design of double-note calls. Those of

Pteropus poliocephalus differ in general structure from their counterparts in other species (Nelson 1964; Gould 1977a).

Not all female bats produce specialized vocalizations when searching for their infants. Female *Rhinolophus ferrumequinum* and *Tadarida condylura* (Matsumura 1979 and Kulzer 1962, respectively) use echolocation calls in this situation. Furthermore, the vespertilionid *Antrozous pallidus* uses directive calls (Fig. 30) when searching for infants; the same vocalization is used in a number of other social settings (Brown 1976).

Although all species of bats appear to have vocalizations which mediate interactions between mothers and young, there is considerable diversity in the way the problems are resolved. In addition to calls produced by isolated young and searching mothers, vocalizations associated with contact between females and their offspring have been reported, although they are usually too soft to have been recorded and described in detail.

A comparison of three vocal repertoires will emphasize the differences in complexity occurring within the Chiroptera. Barclay, Fenton, and Thomas (1979) conducted a field study of the vocal repertoire of *Myotis lucifugus,* a gregarious species often living at high density in roosts, whether night, maternity, or hibernation. They found the repertoire was limited. It included a variety of harsh broadband vocalizations observed in high-density situations in roosts, where individuals were in physical contact with one another. Harsh vocalizations varied in duration according to the ambient temperatures, probably reflecting the bats' body temperatures. Young produced two kinds of isolation calls, longer ones apparently directed at mothers, and shorter ones with a distress call function (Thomson 1980). Females searching for their infants produced double-note calls. During copulation males produced discrete tonal calls, apparently to appease struggling females. No other special vocalizations were associated with copulation and mate selection.

In contrast, Porter (1979a; 1979b) found a larger vocal repertoire in the phyllostomid *Carollia perspicillata.* Harsh broadband vocalizations were common in the repertoire, but did not prevail as they had in *M. lucifugus.* Double-note calls were produced by isolated young and by females searching for their infants. In addition, however, there were several vocalizations associated with male-female interactions in the harem setting, from greeting calls, used by males hovering in front of

A—mating and sex	sexual advertisement		precopulation	copulation	male-female interaction and greeting
	MALE	FEMALE			
Pteropus poliocephalus					
Epomophorus wahlbergi					
Epomops franqueti					
Hypsignathus monstrosus					
Saccopteryx bilineata					
Carollia perspicillata					
Myotis lucifugus					

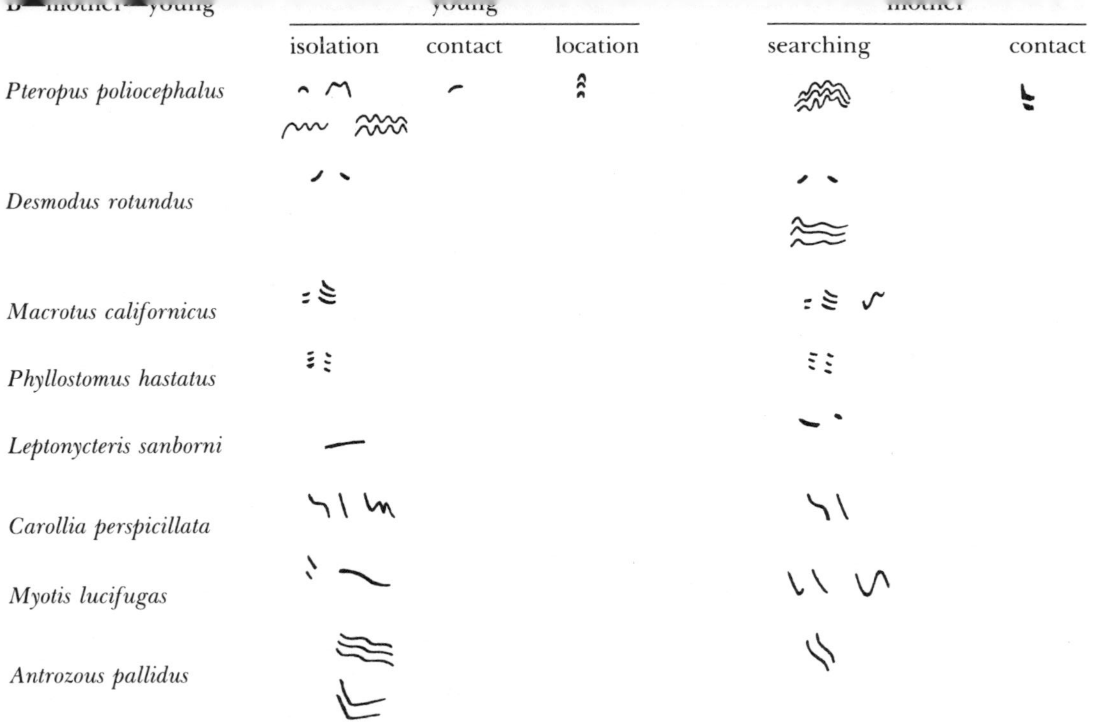

young

| | isolation | contact | location | | searching | contact |

mother

Pteropus poliocephalus

Desmodus rotundus

Macrotus californicus

Phyllostomus hastatus

Leptonycteris sanborni

Carollia perspicillata

Myotis lucifugas

Antrozous pallidus

FIG. 29. A comparison of the frequency-time structures of some vocalizations used in similar situations by different bats. Again, the sketches of the sonograms are not designed to provide detailed information, but rather to permit comparison of basic call designs. Note that tonal vocalizations predominate in both mating and sexual encounters (A) and in mother-young interactions. Under mother-young interactions, it is common for some species to have young producing distinctive isolation calls (e.g., *P. poliocephalus*, *M. lucifugus*, or *A. pallidus*) and females giving distinctive double-note calls. In the phyllostomids, the other species shown in B, double notes are used by isolated infants and by searching mothers. The searching calls of female *A. pallidus* seeking their young are directive calls used in other contexts as well (see Fig. 30 for details).

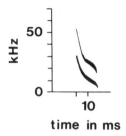

FIG. 30. The directive call of *Antrozous pallidus* adapted from Brown (1976). This vocalization is used by male and female *A. pallidus* in a range of social settings and by females searching for their young.

females, to vocalizations of males during precopulatory behavior. The increased (relative to *M. lucifugus*) repertoire correlates with a more structured social setting and the lack of a copulation call with more elaborate precopulatory displays.

The repertoires of *Pteropus poliocephalus* (Nelson 1964) or *P. giganteus* (Neuweiler 1969) are more complex than either of the two reported above, reflecting, in part, the situation in which the animals roost. A range of vocalizations associated with interactions between females and their young has been identified, including calls produced by isolated young, a category in which Nelson (1964) distinguished several variations. Females searching for their infants produced a distinctive searching call, and both females and young had typical contact calls used after reunions had been achieved. Also included in the repertoires were broadband threat calls, some of which only occurred during interactions between a female and a male attempting to copulate with her. Mating interactions included a wing-flap call produced by males and females (different vocalizations), a precopulatory call, and a copulation call. Males used a distinctive vocalization to advertise their roosting territories, and a similarly designed call in alarm. The calls of males or females when feeding and approached by a conspecific were similar to the alarm and territorial vocalizations. One alarm call was associated with approaches by eagles and resulted in other bats in the camp giving the same call. The eagle alarm call was similar in design to one given by flying bats avoiding collision, although in that context it was not taken up by other bats. Nelson (1964) also reported an orientation vocalization produced by bats en route to their feeding grounds.

The larger repertoires of the two species of *Pteropus* (relative to *M. lucifugus* and *C. perspicillata*) coincide with roosting in the open under relatively stable conditions of social organization and good visibility. *Carollia perspicillata* use darker roosts, which may provide more protection, although in this case comparison of a captive study with a field investigation may have produced the smaller repertoire, and could, for example, account for the absence of alarm calls.

Two important sources of variation in the calls of bats are the notes themselves and the order in which they are juxtaposed. Furthermore, as noted in *P. poliocephalus,* the context in which a call is produced may further influence the message it is transmitting. The most effective demonstration of variability in juxtaposition comes from Bradbury and Emmons's (1974) work on *Saccopteryx bilineata,* where males use long songs and short songs in their interactions.

Another category of bat vocalizations is distress calls. It has long been known to people trying to catch bats that the calls of one individual in a mist net or trap may attract others. Nevertheless, the status of distress calls among bats remains unclear. In a stressful situation most species of bats produce harsh broadband vocalizations, which often attract at least conspecifics. In a set of innovative experiménts, August (1979) demonstrated that some stenodermine phyllostomid bats responded to the real or simulated distress calls of *Artibeus jamaicensis.* His work involved using human whistles to simulate the bat distress calls and manipulation of the recording and playback speeds. The results clearly demonstrated the effect of vocalizations below 20 kHz on the behavior of the bats. Fenton et al. (1976) used playback experiments to demonstrate that within a nursery colony when most females were post-lactating, residents dive-bombed a speaker from which broadband "distress" calls were presented but showed no such response to a variety of other signals. Thomson (1980) observed marked differences in responses of *Myotis lucifugus* to long and short isolation calls in nursery colonies; responses again involved approaches to the speakers. In any of these playback situations the bats responded by approaching, often very closely, the speaker, and in some cases by dive-bombing it. It is possible that these situations simulate conditions when an individual has been captured by a predator and communicates extreme distress, eliciting mobbing behavior, often more effectively from subadults (Smith 1977). Barclay (1982a) also found subadults more responsive to playback presentations of echolocation calls.

Alarm calls and distress calls have been reported from several other bats. The situation in *Pteropus poliocephalus* where the alarm call may be associated with particular dangers (Nelson 1964) may prove analogous to the situation described for some primates which adjust their calls according to the predator (Marler 1983).

The vocalizations of bats cover a broad range of frequencies and serve a variety of functions in orientation and communication. Our knowledge of the calls of bats is limited to those of a few species, and is most complete for calls within our own range of hearing. The responses by bats to playback presentations, now documented in a range of settings (Bradbury and Emmons 1974; Brown 1976; Fenton, Belwood, et al. 1976; Porter 1979a; Barclay, Fenton, and Thomas 1979; Thomson 1980; etc.), clearly indicate the potential for this area of research.

I V

Case Studies

The preceeding information makes it clear that although there are considerable data about communication in the Chiroptera the information is not uniformly distributed and few studies have focused on this topic *per se*. For this reason, among others, the data base is rich in details about one aspect for one species or another, yet totally lacking in comparable areas for other species or other topics for the same species. By presenting the available data for a particular species in the context of its life history we can better appreciate the communication system and, at the same time, highlight some of the important gaps in our knowledge. The purpose of this section is to review the data on behavior and communication for four well-studied species of bats. The species are as representative as any others in a broad sense and they reflect some of the evolutionary variety in the Chiroptera.

Myotis lucifugus, the little brown bat, ranges widely in North America. Adults weigh 8–12 g, depending upon the season, and this has been *the* species for many studies of bats, from morphology to physiology and echolocation (Fenton and Barclay 1980). This is an insectivorous vespertilionid (Microchiroptera), and the data presented below are drawn from a number of sources, principally those cited in Fenton and Barclay (1980) or in Barclay (1982a).

Hypsignathus monstrosus, the hammer-headed bat (males 250–450 g; females 220–375 g), is a pteropodid (Megachiroptera) occurring widely in the high forest and surrounding areas in Africa. The data presented are drawn from the detailed studies of Bradbury (1977b) or D. W. Thomas (1982), with some information also taken from Kingdon (1974).

Pteropus poliocephalus, the gray-headed flying fox, is an Australian pteropodid (500 g) extensively studied by Nelson (1964; 1965). It is frugivorous and provides an interesting comparison with both *H. monstrosus* and with the phyllostomid (Microchiroptera) *Carollia perspicillata.* The latter is a 20 g bat common throughout much of the Neotropics. The data presented here on *C. perspicillata* are drawn mainly from the work of Porter (1978; 1979a; 1979b), Porter and McCracken (1983) and Heithaus and Fleming (1978).

MYOTIS LUCIFUGUS

A midwinter visit to a hibernaculum occupied by *M. lucifugus* will usually give the impression of silence. The bats would normally be torpid, apparently oblivious to their surroundings and to one another. Some individuals hibernate in tightly packed clusters, often composed of hundreds of individuals, while others associate in smaller, loosely packed groups of two to twenty or more bats. Still other individuals hibernate by themselves, out of physical contact with other bats. Disturbance resulting from changes in temperature, light, or sound will induce arousal from torpor, and the silence will be broken by the squawks and squeals of awakened bats and then by the flutter of wings as individuals warm up, take flight, and change locations. Squawks and other vocalizations are usually emitted in response to the jostlings of a neighbor. Some sounds of disturbed bats are audible to the unaided human ear, but many of the vocalizations are ultrasonic. The variety of vocalizations in this setting seems greater than one would encounter in summer roosts, since colder bats produce longer, more drawn out calls at lower rates, but even here the repertoire is limited.

Occasionally from disturbed bats in a hibernaculum one hears copulation calls, apparently emitted by males trying to pacify females with which they are trying to mate. Only struggling females elicit copulation calls, and when males accost other males the vocal responses are usually squawks and squeals. The most parsimonious explanation for attempted homosexual matings during the hibernation period is mistaken identity. Torpid or recently aroused bats may smell less distinctive than active ones.

Our knowledge about the details of behavior during hibernation is very slim. For example, clustering behavior could reflect either a social phenomenon or a limited resource. There is no clear evidence of a thermal benefit accruing to individuals hibernating in clusters. The significance of a predominance of males among large clusters, particularly away from the center of the species' range, is not obvious. In hibernation the communication behavior of *M. lucifugus* is at a low ebb. However, even when torpid the bats respond to stimuli. We know nothing about the behavioral dynamics of hibernation—perhaps not a contradiction in terms.

With spring, defined by the end of the period of occupancy of the hibernacula, there are apparently basic differences in behavior according to sex. We know little about the behavior of males from April to August. Some males leave the hibernaculum after the females leave, but few adult males roost in the nurseries used by the females. We know something about females that establish nursery colonies in buildings, but nothing about either the proportion of the population exploiting artificial roosts or the proportion and behavior of the bats in natural roosts.

Since the arrival of females in the spring is often abrupt, it is tempting to presume that they passed the winter together and left the hibernaculum in a group. This presumption is unsupported by direct evidence, such as sequential observations of females hibernating together and then appearing at a nursery. Furthermore, some nurseries contain females known to hibernate at different sites, so the system is complex. By the time females have reached the nurseries, most have ovulated and are pregnant, fertilized by sperm stored through the winter in the uterus.

Before parturition, levels of interaction among bats in the nurseries are temperature-dependent, with high levels of activity and interaction on hot days, and calm and quiet on cooler days. The vocalizations in these situations are the ubiquitous echolocation calls, as well as squawks and squeals. Females cluster tightly under cool conditions, and then the squawks and squeals reflect changes in posture or disturbances by neighbors. Since the period of gestation is temperature-dependent, individual females appear to benefit from the collective body heat of clusters, but there is no evidence of a strong social structure in these roosts. Sequential observations of tagged, individually recognizable bats provide no evidence of preferred roost-mates, but the few detailed studies *in situ* do not confirm absolutely the lack of social structure. *Myotis lucifugus* also establish small nurseries (less than ten individuals) under natural conditions, but there are no data on interactions and social structure in these settings.

With parturition, the scene changes. The vocal repertoires of the females now include double-note calls (Fig. 28) directed at young and complementing the isolation calls of young separated from their mothers. Retrieval experiments indicate that vocalizations and olfactory cues mediate mother-young recognition. Playback presentations of the isolation calls of young have different effects within colonies. Before parturition there is no response by the females, but afterward females show limited specific responses (investigations), perhaps involving only the mother of the young involved. After some young have started to fly, short isolation calls (Fig. 28) elicit a general aggregation response apparently from other young and reminiscent in a general way of mobbing.

When captured in a mist net or trap, or confined in a cage or sack, *M. lucifugus* produce a range of vocalizations with echolocation calls, squawks, and squeals predominating. These vocalizations attract conspecifics, facilitating the capture of bats in nets or traps. This positive phonotaxis occurs in response to the presentation of echolocation calls, and it seems obvious that active *M. lucifugus* are gregarious, apparently actively seeking conspecifics.

Gregariousness extends to hunting situations, where again the presentation of echolocation calls elicits aggregation behavior. *Myotis lucifugus* frequently hunts in groups, and I have watched up to fifty feeding at one time over a short (10 by 3 m) stretch of stream, apparently exploiting hatches of aquatic insects. Details of possible interactions among feeding individuals are lacking, and in these feeding swarms, the only vocalizations other than echolocation calls are honks (Fig. 18), presumably produced by bats on collision courses.

These bats are also gregarious at night roosts, sites where they congregate to digest food between foraging bouts. The vocalizations from night roosting individuals are echolocation calls, squawks, and squeals, and the animals continue to show positive phonotaxis. Bouts of vocalizations coincide with jostling, usually associated with the arrival of another individual.

The young grow rapidly and can fly by age 18 days. We know nothing about the details of weaning and mother-young interactions during this critical period, but young lose weight during this time and adopt different foraging strategies from those typically used by adults. The young appear to be more responsive than adults to vocalizations of conspecifics during the period after they start to fly.

By early August, *M. lucifugus* start to appear in large numbers at hibernacula, swarming at caves and mines from an hour or so after dark until an hour or so before dawn. The bats appear to feed before visiting the swarming sites. The swarming population includes adults and subadults of both sexes, and partway through August (in Ontario) mating starts at the swarming sites. Within the hibernacula swarming bats are gregarious, usually aggregating around adult males, which avoid one another and are dispersed at sites throughout the cave or mine. Adult males sitting at these sites produce only echolocation calls, but aggregations of active bats generate the usual spectrum of echolocation calls, squawks, and squeals typical of most gatherings. Nasonasal contact is prominent, particularly between adult males and any other bats trying to join their groups. The commencement of mating is marked by the appearance of the copulation calls, which are attractive to other bats. Copulating pairs are often the focus of the attention of many conspecifics, and the antics of the spectators often result in the termination of the coupling.

The mating system appears to be random and promiscuous, as neither sex can protect its investment in mate choice. There is no evidence of differential mating success among males, and males and perhaps females mate with more than one member of the opposite sex. Females, however, visit different males and it is possible that the system is more structured than we suspect.

The mating season starts just before the population of hibernating bats begins to build up, and, as winter approaches, silence again begins to dominate in the hibernacula. Midwinter matings of *M. lucifugus* reported by people banding bats may represent artifacts of disturbance.

Myotis lucifugus is a gregarious species relying on vocalizations as distal cues to locate aggregations of conspecifics. Olfactory cues are also important in interactions, but the repertoire of signals seems quite limited. Many important details are missing from our data base for this species.

Hypsignathus monstrosus

These bats typically pass the day in roosts in foliage, selecting sites relatively high (20–30 m) above the ground and hanging from exposed branches beneath dense umbrellas of vegetation. Roosts may be occupied by solitary bats, or by groups of up to fifteen individuals which, with the exception of mothers and their current young, are not in physical contact with one another. Indeed, roosting individuals maintain a 10 to 15 cm space between them and their nearest neighbor. Observations suggest minimal interactions between roosting individuals, and unless disturbed they appear to sleep, hanging with their noses covered by their wings. Roosting animals are alert, however, and difficult to approach closely. Group size and composition is labile, and these bats frequently switch roost locations. In Ivory Coast, the disturbance associated with capture caused *H. monstrosus* to abandon roosts, which were then not reoccupied within the ensuing fifteen months. In Gabon new roosts were either within 100 m of the old ones, or much further (5 km) away. Spacing of individuals in the roosts seems to be achieved by avoidance rather than by overt communication.

These bats feed largely on figs (*Ficus* spp.), although their diets include a variety of fruits. Several individuals may simultaneously feed in one fig tree, but there are no published observations on the interactions of bats at feeding sites, and no clear evidence of territorial or other agonistic behavior.

The mating sites of *H. monstrosus* are made conspicuous by loud vocalizations of the males displaying on leks. The males are highly specialized for display, and the degree of modification of the sound production system—the larynx fills one-fifth of the body cavity—is among the most spectacular in the animal kingdom. Advertising males roost at about 50 m intervals in linear assemblages along rivers. They call monotonously (Fig. 23), altering their rates of calling when females pass by. Copulations are signaled by release calls given by females, and these three variations in signals permitted Bradbury to assess male success as a function of position in the lek. The precise proximate cues used by females in selecting a mate from among calling males are not known, but the calling obviously functions at least as a distal cue. *Hypsignathus monstrosus* shows exaggeration of calling patterns seen in other species of related bats (Fig. 23).

There are no published data about the interactions of mothers and their young, but females roost with their infants and only mothers and their current young roost in physical contact with one another. Some related species carry their small young during their early days, later presumably leaving them in the day roost or other convenient location when away foraging. Males have nothing to do with the care and rearing of the young.

There is little evidence of a strong social framework in *H. monstrosus*, although longitudinal studies of roosting aggregations might reveal more complex patterns than are currently recognized.

PTEROPUS POLIOCEPHALUS

This species occupies day roosts known as "camps," conspicuous aggregations whose size is influenced by the availability of food. There is seasonal variation in the location of camps and the behavior of the bats therein: in summer camps (September to April or June) the young are born and raised, mates are selected, territories are established and defended, and conception occurs. In winter camps (April or June to September) the sexes are segregated and most of the occupants are young bats in their first year, while most adults are solitary or roost in small, inconspicuous groups.

Between late September and early October females give birth to a single young, and from then until early December the sexes tend to segregate by roosts in the camp, although the overall sex ratio there is about 1:1. Males and females occupy different trees, or one sex may roost higher in a tree than the other. Levels of aggression between individuals are low during this period, but neighbors show a lot of interest in one another, manifested by touching with the thumbs and leaning toward an adjacent bat to sniff its scapular area. Close approaches may elicit agonistic responses, or the two bats may wrap their wings around one another and engage in mutual grooming. Since males often get erections during mutual grooming, it may have a sexual function, but it often (in this season) involves homosexual pairs.

Female *P. poliocephalus* selectively nurse their own young. By age three weeks the young are left in the camps at dusk when females depart to feed. At dusk, females carry their young to parts of the camp with well-foliated trees, often at the edges of the camp, where the impact of roosting bats is less severe. Young are deposited in these peripheral areas and are left by their mothers after being groomed. Before dawn each female returns to the depository and flies about calling to her young (Fig. 28). All of the young reply and the female approaches, eventually landing near one of the calling young. The female sniffs the young's chest and accepts or rejects it. After another session of grooming the accepted young is collected and returned with its mother to the roost site. Vocalizations of young in response to the calls of the female seem to be critical in the reunions, serving as distal cues supplemented by olfactory cues before acceptance or rejection. Although females are selective in the young they will retrieve, young will try to attach themselves to any female. From age three weeks, when they are left in night roosts while their mothers go to forage, young show progressive development, eventually starting to fly by age three months. The first flights involve going from branch to branch, then tree to tree, and progressing to nightly departures from the camp.

The age at weaning varies from four to six months, as does the time when the young leave their mothers to join juvenile packs, which are aggregations composed mainly of subadults. As long as they are with their mothers, young are regularly groomed even when they are capable of doing this for themselves. Large young may also be carried about by their mothers. By late January juvenile packs begin to form when two or three young are found near adults, usually males. The groups of juveniles become larger as the number of family units diminishes, and by late March some of the groups contain fifty juveniles and fifteen adults, ten of which are males. Levels of aggression in juvenile packs are quite low, and normal attacks with the extended thumb claw are pulled or blows delivered with the insides of partly folded wings.

Meanwhile, in December and January, when populations in the summer camps are at their highest, mate selection occurs. During this period when the young are still quite dependent upon their mothers, undisturbed females tend to remain at the same roosts in the camps. In this setting a male will approach a female and her young; the female may respond with wing flaps (of folded wings) and two or three calls (Fig. 28), and the cumulative effect of this reaction may be appeasement to inhibit aggressive and sexual behavior of the male. If the female responds with the wing flap and call, the male turns away and often roosts nearby. If she is passive, however, he will approach and attempt to sniff her scapular area or to lick her genitals. By late January there is close contact between males and females including mutual grooming. From then on males associated with females (one or more) become more and more aggressive. The intensity of male-male fights peaks in late February just before the territory boundaries have been learned by the bats. Disturbance leading to movements within the camp can touch off further intense fighting. Territory boundaries are marked with exudate of the scapular glands.

During March and April territories in summer camps are established and the camp populations are stable, increasing in density from the edges to the center. There are four social groups now: guard groups, family groups, adult groups, and juvenile packs. The guard group contains mainly males, but some females, each with an attendant male, may be included here. These animals on the periphery are those unsuccessful in territorial altercations, and they act as guards, giving alarm calls (Fig. 28) when disturbed. Either acoustical or visual stimuli will elicit alarm responses, and alarms may be given by one of the guards or by any other bat in the camp that detects potential danger.

Family groups (adult male, adult female, and her current young) and adult groups (adult male and one or more adult females) occupy territories where males dominate females and, in polygynous situations, males stop fights between females. Adult group males are more aggressive and mark their territories more often than family group males; adult group males usually hold smaller territories than the family group males. Scent-marking occurs in both groups, where the male hangs and where the female(s) hangs. Group members appear to recognize one another by olfactory cues, usually sniffing the scapular region, but reunions often involve short, soft calls. Family group males occasionally fight with the young, and although the fights are low in intensity and duration, they may contribute to the movement of the young to the juvenile packs.

Males attempt to copulate at other times, but copulation and conception occur in late March when the females are receptive. When mating begins approaches to females are made by males with pinnae directed backward and downward, a "frightened" expression and the antithesis of the aggressive stance. Precopulatory behavior includes mutual grooming and licking of the vulva by the male. Females also may initiate copulation by wrapping their wings and legs around the male; mating is from the rear.

Mated and pregnant females leave territories and aggregate in groups in the camp, signaling the breakup of the territories and commencement of aggregation into sexually segregated groups. The male and female groups then leave the summer camps and move to roosts in different locations. At the end of the summer period sexual segregation, although not absolute, is almost complete. Post-mating aggregations involve bats hanging in close proximity to one another; the bats in these aggregations continue to show great interest in one another and engage in mutual grooming. In winter adults tend to roost alone or in small groups and are rarely encountered in winter camps, which are used mainly by juveniles.

In summer and winter the bats disperse every night to forage. They usually leave the large summer camps in three or four long (over 30 minutes for the column to pass a particular point), wide columns. Several bats may feed in one tree, but each usually defends a feeding territory of about 3 m in each direction. Territory rights are advertised by a feeding cry (Fig. 28), and interactions on feeding grounds include the vocalizations associated with sexual and mother-young behaviors. In this situation, the vocalizations provide a window on the behavior of animals when it is too dark to see them. There are relatively few data on the feeding bats or about possible interactions between individuals on their way to and from feeding grounds.

The summer camps of *P. poliocephalus* may also include roosting *Pteropus scapulatus* and *P. alecto* (= *P. gouldi*). The bats tend to separate by species when population densities are high, but mixed species aggregations occur at low densities during the nonreproductive phase.

Pteropus poliocephalus is a gregarious bat with a clear social structure and a complex repertoire of communication signals.

CAROLLIA PERSPICILLATA

This species typically roosts in groups in sheltered situations, including caves, hollow trees, and buildings, although some individuals may roost alone in foliage. Some adult males roost with groups of females in harem situations, while others form bachelor aggregations, often within the same structure. In a large roost, aggregations of fifty to one hundred *C. perspicillata* are composed of smaller clusters or harems.

Each harem includes one adult male and up to eight females. Other clusters are made up only of males or only of subadults. Some individuals, particularly adult males, show a high degree of site fidelity, while adult females frequently move between groups. Some adult males seem to be more effective at recruiting females than are others. In captivity, harem groups only roost at ceiling level, while other aggregations use lower roost sites as well. Harems occupy the dimmest sites, presumably those farthest from potential disturbance.

Males actively recruit females for their harems through displays which include vocalizations (Fig. 28) and hover flights. When a female that is the object of a male's attention is already in a harem, the harem male responds quickly by shaking his wings and vocalizing to drive away the intruding male. In captivity, newly introduced females usually join existing harems rather than roost with males not holding harems.

Harem males vigorously attempt to keep females within their harems and intruding males out. Scattered harem females are approached and herded into a cluster by the harem male, who uses vocalizations (Fig. 28) and wing-poking to chivy them into position. An intruding male is met at the edge of the cluster of females by the harem male, who uses wing-shaking and vocalizations to discourage the intruder. Sometimes harem females will chase away intruding females, but in general harem females and the harem male interact more with one another than with outsiders of either sex.

Displacement of a harem male by an intruding male occurs after continuous approaches and interactions escalate from threat postures to actual fighting. The displays along this continuum include vocalizations and physical gestures from wing flaps to blows struck with the folded wings and wrists. Harem males and their challengers show dramatically enlarged testes relative to bachelor males. The resident male usually successfully resists attempts to displace him, but in defeat leaves the immediate area of the harem.

Copulations usually occur between harem females and the harem male and are accompanied by vocal displays (Fig. 28), by pursuit of the female by the male, and by olfactory interactions evidenced by the male sniffing the female. Females move often between harems, and the varied paternity of infants born in harems reflects this situation.

Interactions between mothers and their single young are mediated by auditory and olfactory cues. For the first few days, females remain close to and care for their young but leave them alone when away foraging. By age five or six days, an infant may hang out close to but not in physical contact with its mother. Harem males appear to guard young left by their mothers, and youngsters neglected or in danger elicit strong attention from other adults in the harem. Although volant young are attractive to other bats, which will hover in front of them, the young attempt to approach and are retrieved by only their mothers.

Behavioral interactions among harem bats are complex and involve a great deal of communication. Details of the behavior of other groups (bachelors or subadults) are not published, but it seems reasonable to expect the existence of a social hierarchy extending beyond the harems.

Carollia perspicillata feed on the fruits of many species of plants. In Costa Rica, radio-tagged individuals usually move less than 3.2 km from the roost to feed in the wet season, and concentrate their feeding in an area of about 0.06 km². Within this space most individual bats use more than one feeding area, and some visit up to six feeding areas in the space of three or four days. Bats tend to revisit feeding areas used on the previous night, but also add new areas to their selection. Most forage close to the roost ($\bar{x}$ = 0.81 ± 0.52 km), apparently to minimize commuting costs. Flights between feeding areas or between the roosts and feeding areas are direct, and exploratory flights seem uncommon. Experiments using artificial concentrations of fruit, however, suggest that the bats quickly locate and exploit new patches of food. *Carollia perspicillata* usually take food to nearby night roosts for consumption rather than return to the main day roost or feed at the fruiting plant.

Although individual bats showed a high level of overlap in use of the general region around a roost, there is very low overlap in specific feeding areas of tagged individuals. The behavioral dynamics of this avoidance behavior are not known and no obvious (to the observers) communication displays are associated with this pattern of use of space. *Carollia perspicillata* move from roosts to feeding areas through common flight paths, but apparently avoid one another in feeding areas. Only further detailed study will clarify the interactions involved.

This harem-forming species shows a rich communication repertoire and a strongly structured social system.

. . .

This review of the behavior and communication of four species of bats underscores some important similarities and differences. The details sometimes reflect aspects of behavior which are conspicuous, for example, the camps of *P. poliocephalus* or the male displays of *H. monstrosus*. In other cases the ease of keeping the bats in captivity and their suitability for radio-tracking work *(C. perspicillata)* account for the available data, and in the other case, *M. lucifugus*, the data base reflects the local abundance of the bats and their convenience for some kinds of work.

Does the lack of evidence pointing to a strongly structured social system among roosting *M. lucifugus* and *H. monstrosus* reflect reality? I think so, although in the case of *M. lucifugus* observations from small aggregations in natural roosts are essential to put this point in perspective. None of the social interactions conspicuous among roosting *C. perspicillata* and *P. poliocephalus* has been reported from the other two species, and significantly, in *M. lucifugus* and *H. monstrosus* the roosting situations do not involve sexual behavior. The intense echolocation calls of *M. lucifugus* mean greater potential communication among feeding individuals than one would expect in *C. perspicillata*, which uses echolocation calls of low intensity. The two pteropodids do not echolocate.

The association of roosting behavior and sexual activity produces complicated patterns of behavior and rich communication repertoires. Roosting aggregations without sexual behavior seem to reflect the need to remain active and in physical contact with conspecifics *(M. lucifugus)* or the importance of sheltered sites where the animals remain alert and inconspicuous *(H. monstrosus)*.

I think that our knowledge of the biology of these four species is as complete as for any of the Chiroptera, but ongoing or as yet unpublished work promises to change this situation in some bats. The propensity of *Desmodus rotundus* to share blood with individuals who have not fed (e.g., Schmidt 1978; Wilkinson 1983) and that of *Antrozous pallidus* to adjust their positions in clusters to maximize thermal benefit for young (Trune and Slobodchikoff 1976) suggest a strong social fabric in these species. I chose four species to illustrate the completeness and the gaps. Several other species were candidates: *Pteropus giganteus* (Neuweiler 1969), *Antrozous pallidus* (Orr 1954; Brown 1976; Vaughan and O'Shea 1976), *Saccopteryx bilineata* (Bradbury and Emmons 1974), *Desmodus rotundus* (Schmidt 1978; Turner 1975;

Greenhall, Joermann, and Schmidt 1983), or *Artibeus jamaicensis* (Morrison 1978; 1979; 1980; Kunz, August, and Barnett 1983).

Beyond this list of bats that are relatively well studied, there are at least another 840 species. They include some, such as *Eptesicus fuscus, Rhinolophus ferrumequinum,* or *Pipistrellus pipistrellus,* about which we have considerable data in some areas, as well as others known only from a few preserved specimens (e.g., *Harpyionycteris whiteheadi, Haplonycteris fisheri, Depanycteris isabella, Ametrida centurio*). Between the two extremes are many species for which we have only tantalizing scraps of information and species studied in great detail in one or two areas. The majority of species of bats fall into the latter category or must be placed in the one which includes a small smattering of preserved specimens.

V

Communication in the Chiroptera

It is abundantly clear that signals mediating interactions between bats of the same or different species can involve olfactory, visual, and/or auditory media. Although it is tempting to identify vocalizations as being of prime importance to many species of bats, particularly those which echolocate or roost in dark situations, our knowledge of this subject precludes assigning any specific mix to the importance of the different media. Obviously decay features of different signals influence the setting in which they are used. Vocalizations may be effective for long-range signaling, particularly when lighting is poor, but the lack of permanence of the signals means that in other situations scent marks will be more effective. Although scent-marking is involved in many interactions between bats, there is little evidence about the use of scent posts other than the data of Buchler (1980a), which show that they exist in *Myotis lucifugus*.

The singing behavior of *Cardioderma cor* (Vaughan 1976) seems ideally suited to advertising a claim to space. Playback experiments with *Euderma maculatum* (Leonard and Fenton 1984) show that echolocation calls can simultaneously allow a bat to gather information about its surroundings and advertise its presence. Echolocation calls of lower frequency may be more effective in some situations because they carry farther. In *Myotis lucifugus* playback experiments indicate that echolocation calls may simultaneously serve as signals for aggregation (Barclay 1982a).

Scent-marking, often with the exudate of specialized glands, is a prominent feature of the behavior in roosts, in some situations of the space defended in the roost, in other cases of the occupants of a harem. Buchler's (1980a) data on the use of a scent post suggest that marking may involve urine and/or feces as well as glandular products, a situation also reflected in much of the data about interactions between mothers and their young. The tongue-flicking behavior described for *Carollia perspicillata* (Porter 1979a; 1979b) and *Leptonycteris sanborni* (D. J. Howell 1979) in social settings suggests that olfactory cues may be important even without overt scent-marking. In both of these species, the tongue-flicking corresponds with a well-developed Jacobson's organ.

Presentation of spread wings by *Pteropus vampyrus* to conspecifics attempting to land in the feeding tree it occupies (Gould 1977b) is an example of a visual display. Wing displays associated with hovering in *Saccopteryx bilineata* or *Carollia perspicillata,* or the wing-flapping associated with calling by male epomophorine bats are all examples of visual displays accompanied by vocalizations. Other wing displays are associated with aggressive interactions, the thumb commonly being used as a weapon.

Displays involving signals in more than one medium make it difficult to assess which is more (most) important. For example, the data on interactions between mothers and their young in the context of recognition make it clear that olfactory and auditory cues are important. Only in two species is there clear evidence that olfactory cues involve overt scent-marking of the young by its mother (Table 7). The situation is further complicated by the fact that echolocation calls, whether of the mother or the infant, may also play an important role in recognition. The most striking example involves the work of Matsumura (1979; 1981) on *Rhinolophus ferrumequinum* where synchronization of echolocation calls is important in the recognition interaction. Furthermore, Gould's (1983) data on the responses of mothers to abnormally vocalizing (or silent) infants make it clear that vocalizations are important. There remains, however, the fact that sniffing of the young precedes virtually all of the retrievals of young that have been described, and that bats have well-developed spatial memories.

We know enough about the communication behavior of bats to whet the appetite. I suspect that our understanding of the importance of scent posts among bats, if Buchler's (1980a) observations prove generally applicable, will be enhanced as our knowledge of the home ranges of bats improves. Improvements in the miniaturization of telemetry equipment may make this picture easier to understand.

There are a number of phenomena among bats which suggest communication roles which remain unexplored. One of these is the recurring theme of white wings among many species, particularly in Africa. In some cases white or translucent wings contrast with a black body (e.g., the vespertilionid *Eptesicus tennuipinnis*), and in others there is a stripe of white fur between translucent wings and a black body (e.g., the molossid *Tadarida nigeriae*). Contrasting patterns of black and white coloring are spectacular in the African vespertilionids *Glauconycteris egeria* and *G. superba* and in the North American *Euderma maculatum*. Data on the communication behavior of these or other species with contrasting patterns of pelage coloration are totally lacking.

Erectile crests, usually associated with glandular developments between the ears, are common in some African species of molossids. The feature is best developed in *Tadarida chapini,* where the crest is large and conspicuous in males and rudimentary in females or subadult males (Fig. 8). The sexual dimorphism in the feature strongly suggests it plays an important role in interactions between males and females, but again there are no data on the subject.

It is tempting to try to apply sociobiological models to the behavior of bats as it is to mammals in general or indeed to animals as a whole. Hendrichs (1983) discussed some of the drawbacks associated with this approach and stressed that the multidimensional relatedness between different factors governing the behavior of mammals posed a significant drawback to the application of models. He used examples of scent-marking and grooming as patterns of behavior which simultaneously serve a range of functions. Either pattern of behavior can provide, at one time, costs and benefits, not always in convertible currencies. Associated with this problem of multidimensional relatedness is the problem of determining the "original" function of the display or pattern of behavior in question.

Two areas where this situation is particularly relevant to the behavior of bats are feeding and echolocation. Evidence from captive studies with several species shows that bats quickly respond to the sounds of chewing (including chewing by human experimenters) and identify them with a source of food. This easily leads to the suggestion that the information is being pirated. This interpretation is supported by the behavior of bats that avoid divulging these cues when all members of a group are not chewing (e.g., *Nycteris grandis*), and the secrecy of bats such as *Antrozous pallidus* taking prey to a night roost for consumption. There seems to be little doubt that chewing is not intended as a communication signal in the situations for which details are available.

The case involving possible pirating of echolocation calls is less clear. Playback experiments show that the vocalizations which serve one individual in echolocation are exploited by others, as cues for either aggregation or avoidance. From one point of view it is easy to argue that this is not really communication, since the animal producing the calls was not demonstrably trying to send signals. From an evolutionary standpoint, however, there is evidence that echolocation vocalizations originated as social calls (Gould 1983; Fenton 1984), raising the question of identification of the "original" function of the vocalizations. From Hendrichs's (1983) point of view, it probably would not matter, and students of the vocalizations of birds seem to accept the idea that the songs of male passerines (for example) simultaneously serve several functions (e.g., Beer 1975).

I do not see how it will be easy to demonstrate that the bat using echolocation to locate food simultaneously intends to have other bats attend to the signals it produces. In the case of *Myotis lucifugus*, Barclay (1982a) made it clear that the animals producing the calls probably suffered no costs and often received substantial benefit from having other bats aggregate close to them. That *Euderma maculatum* simultaneously find prey and keep conspecifics at a distance where prey appear dispersed means that the individual producing the calls benefits in two ways from the vocalizations it produces.

As Marler (1983) pointed out, it is common to fall into the trap of thinking language when talking about communication, and to rely on language as a model. He warned that this would be safer if we better understood the processes by which language acquires meaning in the course of our own early development. It is clear that bats communicate with one another, but whether or not they have language is another matter.

In several situations there is evidence of a high level of sophistication and individual recognition between bats. The best examples come from interactions between young and adults, and range from clear recognition of mothers by young and young by mothers when it comes time to nurse them, to more complicated situations. I find the results of Porter (1979a; 1979b) most exciting in this regard. She found that not only did female *Carollia perspicillata* recognize their own young, but that females were associated with their babies by other members of the harem in which they roosted, including the harem male. That female harem-mates and the harem male harassed a female that was neglecting her infant tells us a great deal about the bats' view of their social setting.

The situation is made even more interesting by the recent description of the role played by genetics. Porter and McCracken (1983) used electrophoretic data to demonstrate that about 25 percent of the time male *C. perspicillata* attend to young which cannot (genetically) be theirs. The presence of strong social units is common among many bats; at this time most of our data come from the Phyllostomidae, but I suspect that we will find that the situation in *C. perspicillata* is not exceptional. It may be of equal importance that the behavior of males does not necessarily follow the predictions of some sociobiological theory, harping back to the discussion of Hendrichs (1983).

When the data on the social integrity of some groups of bats are considered in the light of the recent evidence for observational learning (Gaudet and Fenton 1984), there is clearly the potential for very important interactions which hinge on communication. I think it significant that the observational learning described by Gaudet and Fenton (1984) involved two species that are relatively asocial, albeit gregarious (*Myotis lucifugus* and *Eptesicus fuscus*), and the more social *Antrozous pallidus*. Gaudet and Fenton did not identify the stimuli most important in mediation of observational learning, but the general behavior of the bats, from chewing to echolocation and patterns of flight, could have been involved.

The combination of individual recognition, extending to members of the immediate group, and the potential for observational learning clearly establishes the situation which Hendrichs (1983) identifies as one where "breakthroughs" could occur.

The use of similar vocalizations in different contexts by *Pteropus poliocephalus* and *P. giganteus*, eliciting different responses from conspecifics in the area, further suggests a sophisticated system of communication. A parallel example, admittedly of less impact, is the use that *Antrozous pallidus* make of directive calls, relying on these vocalizations in interactions between mothers and their babies, and between adults in a variety of situations (Brown 1976; Vaughan and O'Shea 1976). In the context of roosting, *A. pallidus* obtain clear physiological benefits from association with conspecifics (Trune and Slobodchikoff 1976).

Observations of this nature, particularly those involving alarm calls, are reminiscent of the work on monkey vocalizations, which Marler (1983) suggested were evidence of special processing of some signals. Marler pointed out that this proposal is an echo of the idea of innate releasers, which could be taken as supporting the view that animals are instinctional machines. Innate responsiveness could be the result of genetic guidelines for learning how to process important stimuli. It therefore could be associated with flexibility rather than being the result of a mechanical, innate response which carries with it an implied stigma. Marler (1983) preferred the interpretation which invoked flexibility. Flexibility appears to be a feature of the behavior of bats and an important one in the context of their exploitation of a range of opportunities.

The communication behavior of bats clearly offers biologists an excellent field in which to test many of the ideas central to different areas of biology. The richness of the opportunities derives in part from the diversity of the Chiroptera. Within one order there is a spectrum of social settings from solitary to gregarious to social. Superimposed on this spectrum is a range of feeding habits, from species relying mainly on plant products to those feeding only on other animals. This situation is in turn set against an evolutionary background that offers good opportunities for comparison of closely or more distantly related forms occupying similar settings in different parts of the world.

The technological tools currently available make it possible to collect much more data on the behavior of bats despite their nocturnal habits and proclivity for roosting in out-of-the-way places. The echolocation of many species provides a convenient window on their activities and adds another dimension to studies of communication.

We are a long way from the last word on communication in bats. Like other animals for which there are data, bats rely on whatever stimuli are available to them to collect information about their surroundings and about other animals. The species which are social, more than those which are gregarious, clearly offer the best ground for studying communication and behavior.

REFERENCES

Advani, R. 1981. Seasonal fluctuations in the feeding ecology of the Indian false vampire, *Megaderma lyra lyra* (Chiroptera: Megadermatidae) in Rajastan. *Z. Saugetierk.*, 46:90–93

Allen, G. M. 1939. *Bats.* Cambridge: Harvard Univ. Press.

Altenbach, J. S. 1979. Locomotor morphology of the vampire bat, *Desmodus rotundus.* Spec. Pub. no. 6. Stillwater, Oklahoma: Am. Soc. Mammalogists.

Anand, T. C. 1965. Reproduction in the rat-tailed bat, *Rhinopoma kinneari. J. Zool. London,* 147:147–155.

Anthony, E. L. P., and T. H. Kunz. 1977. Feeding strategies of the little brown bat, *Myotis lucifugus,* in southern New Hampshire. *Ecology,* 58:775–786.

Anthony, E. L. P., M. H. Stack, and T. H. Kunz. 1981. Night roosting and the nocturnal time budget of the little brown bat, *Myotis lucifugus:* effects of reproductive status, prey density, and environmental conditions. *Oecologia,* 51:151–156.

Archer, M. 1978. Australia's oldest bat, a possible rhinolophid. *Proc. R. Soc. Queensland,* 89:23.

Armstrong, R. B., C. D. Ianuzzo, and T. H. Kunz. 1977. Histochemical and biochemical properties of flight muscle fibres in the little brown bat, *Myotis lucifugus. J. Comp. Physiol.,* 119:141–154.

Asdell, S. A. 1964. Chiroptera. In *Patterns of Mammalian Reproduction,* pp. 64–122. Ithaca: Cornell Univ. Press.

August, P. X. V. 1979. Distress calls in *Artibeus jamaicensis:* ecology and evolutionary implications. In *Vertebrate Ecology in the Northern Neotropics,* ed. J. F. Eisenberg, pp. 151–160. Washington: Smithsonian Inst.

Ayala, S. C., and A. D'Alessandro. 1973. Insect feeding behavior of some Colombian fruit-eating bats. *J. Mamm.,* 54:266–267.

Ayensu, E. S. 1974. Plant and bat interactions in west Africa. *Ann. Missouri Bot. Gard.,* 61:702–727.

137

Baker, R. J. 1981. Chromosome flow between chromosomally characterized taxa of a volant mammal, *Uroderma bilobatum* (Chiroptera: Phyllostomidae). *Evolution*, 35:296–305.

Baker, R. J., W. J. Bleier, and W. R. Atchley. 1975. A contact zone between karyotypically characterized taxa of *Uroderma bilobatum* (Mammalia: Chiroptera). *Syst. Zool.*, 24:133–142.

Baker, R. J., J. K. Jones, Jr., and D. C. Carter (eds.). 1976. Biology of Bats of the New World Family Phyllostomatidae. Part I. Lubbock: Special Publication, The Museum, Texas Tech. University.

————1977. Biology of Bats of the New World Family Phyllostomatidae. Part II. Lubbock: Special Publication, The Museum, Texas Tech. University.

————1979. Biology of Bats of the New World Family Phyllostomatidae. Part III. Lubbock: Special Publication, The Museum, Texas Tech. University.

Barbour, R. W. and W. H. Davis. 1969. *Bats of America*. Lexington: Univ. of Kentucky Press.

Barclay, R. M. R. 1982a. Interindividual use of echolocation calls: eavesdropping by bats. *Behav. Ecol. Sociobiol.*, 10:271–275.

————1982b. Night roosting behavior of the little brown bat, *Myotis lucifugus*. *J. Mamm.*, 63:464–474.

————1982c. Foraging strategies of lasiurines at Delta, Manitoba. *Bat Research News*, 23:59.

————1983. Echolocation calls of emballonurid bats from Panama. *J. Comp. Physiol.*, 151:515–520.

Barclay, R. M. R., M. B. Fenton, and D. W. Thomas. 1979. Social behavior of the little brown bat, *Myotis lucifugus*. II. Vocal communication. *Behav. Ecol. Sociobiol.*, 6:137–146.

Barclay, R. M. R., M. B. Fenton, M. D. Tuttle, and M. J. Ryan. 1981. Echolocation calls produced by *Trachops cirrhosus* (Chiroptera: Phyllostomatidae) hunting for frogs. *Can. J. Zool.* 59:750–753.

Barclay, R. M. R., and D. W. Thomas. 1979. Copulation calls of *Myotis lucifugus*: a discrete situation-specific communication signal. *J. Mamm.*, 60:632–634.

Beck, A. J., and Lim Boo-Liat. 1973. Reproductive biology of *Eonycteris spelaea* Dobson (Megachiroptera) in West Malaysia. *Acta Trop.*, 30:251–260.

Beer, C. G. 1975. Multiple functions and gull displays. In *Function and Evolution in Behavior, Essays in Honour of Professor Niko Tinbergen*, ed. G. Baerends, C. Neer, and A. Manning, pp. 16–54. Oxford: Clarendon Press.

Belknap, D. B., and R. A. Suthers. 1982. Brainstem auditory evoked responses to tone bursts in the echolocating bat, *Rousettus*. *J. Comp. Physiol.*, 146:283–289.

Bell, G. P. 1980a. Habitat use and responses to patches of prey by desert insectivorous bats. *Can. J. Zool.*, 58:1876–1883.

————1980b. A possible case of interspecific transmission of rabies in insectivorous bats. *J. Mamm.*, 61:528–530.

————1982a. Behavioral and ecological aspects of gleaning by a desert insectivorous bat, *Antrozous pallidus* (Chiroptera: Vespertilionidae). *Behav. Ecol. Sociobiol.*, 10:217–223.

————1982b. Prey location and sensory ecology of two species of gleaning, insectivorous bats, *Antrozous pallidus* (Vespertilionidae) and *Macrotus californicus* (Phyllostomatidae). Ph.D. Thesis, Carleton University, Ottawa, Canada.

Bell, G. P., and M. B. Fenton. 1984. The use of Doppler-shifted echoes as a flutter detection and clutter rejection system: the echolocation and feeding behavior of *Hipposideros ruber* (Chiroptera: Hipposideridae). *Behav. Ecol. Sociobiol.*

Belwood, J. J. 1982. Foraging in the Hawaiian hoary bat, *Lasiurus cinereus*. *Bat Research News*, 23:60.

Bernard, R. T. F. 1980. Female reproduction in five species of Natal cave-dwelling Microchiroptera. Ph.D. Dissertation, University of Natal, Pietermaritzburg, South Africa.

Bhatnagar, K. P. 1980. The chiropteran vomeronasal organ: its relevance to the phylogeny of bats. In *Proceedings of the Fifth International Bat Research Conference*, ed. D. E. Wilson and A. L. Gardner, pp. 289–316. Lubbock: Texas Tech. Press.

Bhatnagar, K. P., and F. C. Kallen. 1974a. Cribriform plate of ethmoid, olfactory bulb and olfactory acuity in forty species of bats. *J. Morph.*, 142:71–90.

———1974b. Morphology of the nasal cavities and associated structures in *Artibeus jamaicensis* and *Myotis lucifugus*. *Am. J. Anat.*, 139:167–189.

———1975. Quantitative observations on the nasal epithelia and olfactory innervation in bats. *Acta Anat.*, 91:272–282.

Birney, E. C., and R. M. Timm. 1975. Dental ontogeny and adaptation in *Diphylla ecaudata*. *J. Mamm.*, 56:204–207.

Black, H. L. 1972. Differential exploitation of moths by the bats *Eptesicus fuscus* and *Lasiurus cinereus*. *J. Mamm.*, 53:598–601.

Bleier, W. J. 1975. Early embryology and implantation in the California leaf-nosed bat *Macrotus californicus*. *Anat. Rec.*, 182:237–254.

Bogan, M. A. 1972. Observations on parturition and development in the hoary bat, *Lasiurus cinereus*. *J. Mamm.*, 53:611–614.

Bower, S. M., and P. T. K. Woo. 1981a. Two new species of trypanosomes (subgenus *Schizotrypanum*) in bats from southern Ontario. *Can. J. Zool.*, 59:530–545.

———1981b. Development of *Trypanosoma (Schizotrypanum) hedricki* in *Cimex brevis* (Hemiptera: Cimicidae). *Can. J. Zool.*, 59:546–554.

Bradbury, J. W. 1977a. Social organization and communication. In *Biology of Bats*, vol. 3, ed. W. A. Wimsatt, pp. 1–72. New York: Academic Press.

———1977b. Lek mating behavior in the hammer-headed bat. *Z. Tierpsychol.*, 45:225–255.

———1981. The evolution of leks. In *Natural Selection and Social Behavior*, ed. R. D. Alexander and D. W. Tinkle, pp. 138–169. New York: Chiron Press.

Bradbury, J. W., and L. H. Emmons. 1974. Social organization in some Trinidad bats. I. Emballonuridae. *Z. Tierpsychol.*, 36:137–183.

Bradbury, J. W., D. Morrison, E. Stashko, and R. Heithaus. 1979. Radio-tracking methods for bats. *Bat Research News*, 30:9–17.

Bradbury, J. W., and F. Nottebohm. 1969. The use of vision by little brown bats, *Myotis lucifugus* under controlled conditions. *Anim. Behav.*, 17:480–485.

Bradbury, J. W., and S. L. Vehrencamp. 1976. Social organization and foraging in emballonurid bats. I. Field studies. *Behav. Ecol. Sociobiol.*, 1:337–381.

———1977. Social organization and foraging in emballonurid bats. IV. Parental investment strategies. *Behav. Ecol. Sociobiol.*, 2:19–29.

Bradshaw, G. V. R. 1961. Le cycle de reproduction de *Macrotus californicus* (Chiroptera: Phyllostomatidae). *Mammalia*, 25:117–119.

————1962. Reproductive cycle of the California leaf-nosed bat, *Macrotus californicus. Science,* 136:645–646.

Bronson, F. H. 1983. Chemical communication in house mice and deer mice: functional roles in reproduction of wild populations. In *Advances in the Study of Mammalian Behavior,* ed. J. F. Eisenberg and D. G. Kleiman, pp. 198–238. *Am. Soc. Mamm.,* Spec. Pub. no. 7.

Brosset, A. 1962. The bats of central and western India, pt.3. *J. Bombay Nat. Hist. Soc.,* 59:707–746.

————1966. *La Biologie des Chiroptères.* Paris: Masson et Cie.

Brosset, A., and C. D. Deboutteville. 1966. Le regime alimentaire du vespertilion de Daubenton, *Myotis daubentoni. Mammalia,* 30:247–251.

Brown, P. 1976. Vocal communication in the pallid bat, *Antrozous pallidus. Z. Tierpsychol.,* 41:34–54.

Brown, P. E., T. W. Brown, and A. D. Grinnell. 1983. Echolocation, development, and vocal communication in the lesser bulldog bat, *Noctilio albiventris. Behav. Ecol. Sociobiol.,* 13:287–298.

Brown, P. E., A. D. Grinnell, and J. B. Harrison. 1978. The development of hearing in the pallid bat, *Antrozous pallidus. J. Comp. Physiol.,* 126:169–182.

Buchler, E. R. 1976. A chemiluminescent tag for tracking bats and other small, nocturnal animals. *J. Mamm.,* 57:173–176.

————1980a. The development of flight, foraging, and echolocation in the little brown bat *(Myotis lucifugus). Behav. Ecol. Sociobiol.,* 6:211–218.

————1980b. Evidence for the use of scent post by *Myotis lucifugus. J. Mammal.,* 61:525–528.

Buchler, E. R., and S. B. Childs. 1981. Orientation to distant sounds by foraging big brown bats *(Eptesicus fuscus). Anim. Behav.,* 29:428–432.

Burnett, C. D., and T. H. Kunz. 1982. Growth rates and age-estimation in *Eptesicus fuscus* and comparison with *Myotis lucifugus. J. Mamm.,* 63:33–41.

Busnel, R-G., and J. F. Fish (eds.) 1980. *Animal Sonar Systems.* New York: NATO Advanced Study Institute A28, Plenum Press.

Chase, J. 1972. The role of vision in echolocating bats. Ph.D. Thesis, Indiana University, Bloomington.

Childs, S. B., and E. R. Buchler. 1981. Perception of simulated stars by *Eptesicus fuscus* (Vespertilionidae): a potential navigational mechanism. *Anim. Behav.,* 29:1028–1035.

Christian, J. J. 1956. The natural history of a summer aggregation of the big brown bat, *Eptesicus fuscus fuscus. Am. Midl. Nat.,* 55:66–95.

Constantine, D. G. 1966. Ecological observations on lasiurine bats in Iowa. *J. Mamm.,* 47:34–41.

Cooper, J. G., and K. P. Bhatnagar. 1976. Comparative anatomy of the vomeronansal organ complex in bats. *J. Anat.,* 122:571–601.

Courrier, R. 1927. Etude sur le determinisme des caracteres sexuels secondaires chez quelques mammiferes a l'activite testiculaire periodique. *Arch. Biol.,* 37:173–334.

Dalland, J. I. 1965. Hearing sensitivity in bats. *Science,* 150:1185–1186.

Daniel, M. J. 1979. The New Zealand short-tailed bat, *Mystacina tuberculata;* a review of present knowledge. *New Zealand J. Zool.,* 6:357–370.

Dapson, R. W., E. H. Studier, M. J. Buckingham, and A. L. Studier. 1977. Histochemistry of odoriferous secretions from integumentary glands of three species of bats. *J. Mamm.*, 58:531–535.

Davis, R. B. 1969. Growth and development of young pallid bats, *Antrozous pallidus*. *J. Mamm.*, 50:729–736.

Davis, R. B., C. F. Herreid II, and H. L. Short. 1962. Mexican free-tailed bats in Texas. *Ecol. Monog.*, 32:311–346.

Davis, W. B. 1944. Notes on Mexican mammals. *J. Mamm.*, 25:370–403.

Davis, W. H., R. B. Barbour, and M. D. Hassell. 1968. Colonial behavior of *Eptesicus fuscus*. *J. Mamm.*, 49:44–50.

Di Maio, F. H. P., and J. Tonndorf. 1978. The terminal zone of the external auditory meatus in a variety of mammals. *Arch. Otolaryngol.*, 104:570–575.

Dinale, G. 1964. Studi sui Chirotteri italiani. II. Il raggiungimento della maturita sessuale in *Rhinolophus ferrumequinum*, Schreber. *Atti Soc. Ital. Sci. Nat. Mus. Civ. Stor. Nat. Milano*, 103:141–153.

———1968. Studi sui Chirotteri italiani. VII. Sul raggiungimento della maturita sessuale dei Chirotteri europei: ed in particolare nei Rhinolophidae. *Arch. Zool. Ital.*, 53:51–71.

Douglas, A. M. 1967. The natural history of the ghost bat, *Macroderma gigas* (Microchiroptera: Megadermatidae) in western Australia. *West. Australian Nat.*, 10:125–137.

Downes, W. L., Jr. 1964. Unusual roosting behavior in red bats. *J. Mamm.*, 45:143–144.

Dwyer, P. D. 1963. The breeding biology of *Miniopterus schreibersii blepotis* (Temminck) (Chiroptera) in north-eastern New South Wales. *Aust. J. Zool.*, 11:219–240.

———1968. The biology, origin, and adaptation of *Miniopterus australis* (Chiroptera) in New South Wales. *Aust. J. Zool.*, 16:49–68.

———1970. Foraging behavior of the Australian large-footed *Myotis* (Chiroptera). *Mammalia*, 34:76–80.

———1975. Notes on *Dobsonia moluccensis* (Chiroptera) in the New Guinea highlands. *Mammalia*, 39:113–118.

Edgerton, H. E., P. F. Spangle, and J. K. Baker. 1966. Mexican free-tail bats: photography. *Science*, 153:201–203.

Eisenberg, J. F. 1981. *The Mammalian Radiations, an Analysis of Trends in Evolution, Adaptation, and Behavior*. Chicago: Univ. of Chicago Press.

Eisenberg, J. F., and D. G. Kleiman. 1972. Olfactory communication in mammals. *Ann. Rev. Ecol. Syst.*, 3:1–32.

Erkert, H. G. 1982. Ecological aspects of bat activity rhythms. In *Ecology of Bats*, ed. T. H. Kunz, pp. 201–242. New York: Plenum Press.

Ewer, R. F. 1968. *Ethology of Mammals*. London: Elek Science.

Fattu, J. M., and R. A. Suthers. 1981. Subglottic pressure and the control of phonation by the echolocating bat, *Eptesicus*. *J. Comp. Physiol.*, 143:465–475.

Fayenuwo, J. O., and L. B. Halstead. 1974. Breeding cycle of straw-colored fruit bat, *Eidolon helvum*, at Ife Ife, Nigeria. *J. Mamm.*, 55:453–454.

Feduccia, A. 1980. *The Age of Birds.* Cambridge: Harvard Univ. Press.

Fenton, M. B. 1970. The deciduous dentition and its replacement in *Myotis lucifugus* (Chiroptera: Vespertilionidae). *Can. J. Zool.,* 48:817–820.

———1977. Variation in the social calls of little brown bats *(Myotis lucifugus). Can. J. Zool.,* 55:1151–1156.

———1980. Adaptiveness and ecology of echolocation in terrestrial (aerial) systems. In *Animal Sonar Systems,* ed. R-G. Busnel and J. F. Fish, pp. 427–446. New York: NATO Advanced Study Institutes A28, Plenum Press.

———1982a. Echolocation, insect hearing, and the feeding ecology of insectivorous bats. In *Ecology of Bats,* ed. T. H. Kunz, pp. 261–285. New York: Plenum Press.

———1982b. Echolocation calls and patterns of hunting and habitat use of bats (Microchiroptera) from Chillagoe, North Queensland. *Aust. J. Zool.,* 30:417–425.

———1983a. *Just Bats.* Toronto: Univ. of Toronto Press.

———1983b. Roosts used by the African bat *Scotophilus leucogaster* (Chiroptera: Vespertilionidae). *Biotropica.,* 15:129–132.

———1984. Echolocation; implications for ecology and evolution of bats. *Quart. Rev. Bio.,* 59:33–53.

———in press. Sperm competition? The case of vespertilionid and rhinolophid bats. In *Sperm Competition and the Evolution of Animal Mating Systems,* ed. R. L. Smith, pp. 573–587. New York: Academic Press.

Fenton, M. B., and R. M. R. Barclay. 1980. *Myotis lucifugus. Mammalian Species,* no. 142:1–8.

Fenton, M. B., and G. P. Bell. 1979. Echolocation and feeding behaviour in four species of *Myotis* (Chiroptera). *Can. J. Zool.,* 57:1271–1277.

———1981. Recognition of species of insectivorous bats by their echolocation calls. *J. Mamm.,* 62:233–243.

Fenton, M. B., G. P. Bell, and D. W. Thomas. 1981. Echolocation and feeding behaviour of *Taphozous mauritianus* (Chiroptera: Emballonuridae). *Can. J. Zool.,* 58:1774–1777.

Fenton, M. B., J. J. Belwood, J. H. Fullard, and T. H. Kunz. 1976. Responses of *Myotis lucifugus* (Chiroptera: Vespertilionidae) to calls of conspecifics and to other sounds. *Can. J. Zool.,* 54:1443–1448.

Fenton, M. B., N. G. H. Boyle, T. M. Harrison, and D. J. Oxley. 1977. Activity patterns, habitat use, and prey selection by some African insectivorous bats. *Biotropica,* 9:73–85.

Fenton, M. B., and T. H. Fleming. 1976. Ecological interactions between bats and nocturnal birds. *Biotropica,* 8:104–110.

Fenton, M. B., and J. H. Fullard. 1981. Moth hearing and the feeding strategies of bats. *Am. Sci.,* 69:266–275.

Fenton, M. B., C. L. Gaudet, and M. L. Leonard. 1983. Feeding behaviour of the bats *Nycteris grandis* and *Nycteris thebaica* (Nycteridae) in captivity. *J. Zool. Lond.*

Fenton, M. B., H. G. Merriam, and G. L. Holroyd. 1983. Bats of Kootenay, Glacier, and Mount Revelstoke national parks in Canada: identification by echolocation calls, distribution and biology. *Can. J. Zool.,* 61:2503–2508.

Fenton, M. B., and G. K. Morris. 1976. Opportunistic feeding by desert bats *(Myotis* spp.). *Can. J. Zool.,* 54:526–530.

Fenton, M. B., and D. W. Thomas. In press. Migrations and dispersal of bats (Chiroptera). Proc. of Conf. on Migrations. Texas Univ.

Fenton, M. B., D. W. Thomas, and R. Sasseen. 1981. *Nycteris grandis* (Nycteridae): an African carnivorous bat. *J. Zool. Lond.*, 194:461–465.

Fiedler, J. 1979. Prey catching with and without echolocation in the Indian false vampire *(Megaderma lyra)*. *Behav. Ecol. Sociobiol.*, 6:155–160.

Findley, J. S. 1976. The structure of bat communities. *Am. Nat.*, 110:129–139.

Findley, J. S., and D. E. Wilson. 1974. Observations on the neotropical diskwinged bat, *Thyroptera tricolor* Spix. *J. Mamm.*, 55:562–571.

Fleming, T. H. 1971. *Artibeus jamaicensis:* delayed embryonic development in a neotropical bat. *Science*, 171:402–404.

———1973. The reproductive cycles of three species of opossums and other mammals in the Panama Canal Zone. *J. Mamm.*, 54:439–455.

———1982. Foraging strategies of plant-visiting bats. In *Ecology of Bats*, ed. T. H. Kunz, pp. 287–325. New York: Plenum Press.

Fleming, T. H., E. T. Hooper, and D. E. Wilson. 1972. Three Central American bat communities: structure, reproductive cycles, and movement patterns. *Ecology*, 53:555–569.

Forman, G. L., R. J. Baker, and J. D. Gerber. 1968. Comments on the systematic status of vampire bats (Desmodontidae). *Syst. Zool.*, 17:417–425.

Fowler, B. C., and J. M. Zook. 1982. Electrophysiological analysis of the chiropteran somatosensory cortex. *Bat Research News*, 23:69.

Fullard, J. H. 1982a. Echolocation assemblages and their effects on moth auditory systems. *Can. J. Zool.*, 60:2572–2576.

———1982b. Echolocatory and agonistic vocalizations of the Hawaiian hoary bat, *Lasiurus cinerius. Bat Research News*, 23:70.

Fullard, J. H., M. B. Fenton, and C. L. Furlonger. 1983. Sensory relationships of moths and bats sampled from two nearctic sites. *Can J. Zool.*, 61:1752–1757.

Fullard, J. H., M. B. Fenton, and J. A. Simmons. 1979. Jamming bat echolocation: the clicks of arctiid moths. *Can. J. Zool.*, 57:657–649.

Fullard, J. H., and D. W. Thomas. 1981. Detection of certain African insectivorous bats by sympatric, tympanate moths. *J. Comp. Physiol.*, 143:363–368.

Gaisler, J. 1965. The female sexual cycle and reproduction in the lesser horseshoe bat *(Rhinolophus hipposideros hipposideros* Bechstein, 1800). *Acta Soc. Zool. Bohem.*, 29:336–352.

Gaisler, J., V. Hanak, and J. Dungel. 1979. A contribution to the population ecology of *Nyctalus noctula* (Mammalia, Chiroptera). *Acta Sci. Nat. Brno.*, 13:1–38.

Gaisler, J., and M. Titlbach. 1964. The male sexual cycle in the lesser horseshoe bat *(Rhinolophus hipposideros hipposideros* Bechstein, 1800). *Acta Soc. Zool. Bohem.*, 28:268–277.

Galef, B. G., Jr. 1976. Social transmission of acquired behavior: a discussion of tradition and social learning in vertebrates. In *Advances in the Study of Behavior*, vol. 6, ed. S. Rosenblatt, R. A. Hinde, E. Shaw, and C. Beer, pp. 77–100. New York: Academic Press.

Gardner, A. L. 1977. Feeding habits. In *Biology of Bats of the New World Family Phyllostomatidae*, Part II, ed. R. J. Baker, J. K. Jones, Jr., and D. C. Carter, pp. 293–350. Lubbock: Special Pub. The Museum, Texas Tech. Univ.

Gates, W. H. 1936. Keeping bats in captivity. *J. Mamm.*, 17:268–273.

Gaudet, C. L. 1982. Behavioural basis of foraging flexibility in three species of insectivorous bats; an experimental study using captive *Antrozous pallidus, Eptesicus fuscus* and *Myotis lucifigus*. M.Sc. Thesis, Department of Biology, Carleton University, Ottawa, Canada.

Gaudet, C. L., and M. B. Fenton. 1984. Observational learning in three species of insectivorous bats (Chiroptera). *Anim. Behav.*

Geggie, J. F. 1983. The effects of urbanization on habitat use by the big brown bat, *Eptesicus fuscus*. M.Sc. Thesis, Department of Biology, Carleton University, Ottawa, Canada.

Gillette, D. D. 1975. Evolution of feeding strategies in bats. *Tebiwa*, 18:39–48.

Gingerich, P. D., N. A. Wells, D. E. Russell, and S. M. Ibrahim Shah. 1983. Origin of whales in epicontinental remnant seas: new evidence from the early Eocene of Pakistan. *Science*, 220:403–406.

Glickstein, M., and M. Millodot. 1970. Retinoscopy and eye size. *Science*, 168:605–606.

Goodwin, R. E. 1970. Ecology of Jamaican bats. *J. Mamm.*, 51:571–579.

Gopalakrishna, A. 1947. Studies on the embryology of Microchiroptera. 1. Reproduction and breeding seasons in the south Indian vespertilionid bat, *Scotophilus wroughtoni* (C. Thomas). *Proc. Indian Acad. Sci.*, 26:219–232.

Gopalakrishna, A., and A. Madhavan. 1971. Survival of spermatozoa in the female genital tract of the Indian vespertilionid bat, *Pipistrellus ceylonicus chrysothrix* (Wroughton). *Proc. Indian Acad. Sci. B.*, 73:43–49.

————1978. Viability of inseminated spermatozoa in the Indian vespertilionid bat *Scotophilus heathi* (Horsefield). *Indian J. Exp. Biol.*, 16:852–854.

Gould, E. 1971. Studies of maternal-infant communication and development of vocalizations in the bats *Myotis* and *Eptesicus*. *Comm. Behav. Biol.*, 5:263–313.

————1977a. Echolocation and communication. In *Biology of Bats of the New World Family Phyllostomatidae*, Part II, ed. R. J. Baker, J. K. Jones, Jr., and D. C. Carter, pp. 247–279. Lubbock: Texas Tech. Univ. Press.

————1977b. Foraging behavior of *Pteropus vampyrus* on the flowers of *Durio zibethinus*. *Malay. Nat. J.*, 30:53–57.

————1978a. Foraging behavior of Malaysian nectar-feeding bats. *Biotropica*, 10:184–193.

————1978b. Opportunistic feeding by tropical bats. *Biotropica*, 10:75–76.

————1980. Vocalizations of Malaysian bats (Microchiroptera and Megachiroptera). In *Animal Sonar Systems*, ed. R-G. Busnel and J. F. Fish, pp. 901–904. New York: NATO Advanced Study Institutes A28, Plenum Press.

————1983. Mechanisms of mammalian auditory communication. In *Advances in the Study of Mammalian Behavior*, ed. J. F. Eisenberg and D. G. Kleiman, pp. 265–342. Am. Soc. Mammalogists, Spec. Pub., no. 7.

Green, R. A. 1965. Observations on the little brown bat, *Eptesicus pumilus* (Gray) in Tasmania. *Rec. Queen Victoria Mus.*, 10:1–16.

Greenhall, A. M. 1972. The biting and feeding habits of the vampire bat, *Desmodus rotundus*. *J. Zool. Lond.*, 168:451–461.

Greenhall, A. M., G. Joermann, and U. Schmidt. 1983. *Desmodus rotundus. Mammalian Species*, 202:1–6.

Griffin, D. R. 1958. *Listening in the Dark*. New Haven: Yale Univ. Press.

———1970. Migrations and homing of bats. In *Biology of Bats*, vol. I, ed. W. A. Wimsatt, pp. 233–264. New York: Academic Press.

———1971. The importance of atmospheric attenuation for the echolocation of bats (Chiroptera). *Anim. Behav.*, 19:55–61.

———1980. Early history of research on echolocation. In *Animal Sonar Systems*, ed. R-G. Busnel and J. F. Fish, pp. 1–10. New York: Plenum Press.

Griffin, D. R., and A. Novick. 1955. Acoustic orientation of neotropical bats. *J. Exp. Zool.*, 130:251–300.

Griffin, D. R., and D. Thompson. 1982. High altitude echolocation of insects by bats. *Behav. Ecol. Sociobiol.*, 10:303–306.

Griffin, D. R., F. A. Webster, and C. R. Michael. 1960. The echolocation of flying insects by bats. *Anim. Behav.*, 8:141–154.

Grinnell, A. D. 1980. Dedication. In *Animal Sonar Systems*, ed. R-G. Busnel and J. F. Fish, pp. xix–xxiv. New York: NATO Advanced Study Institutes A28, Plenum Press.

Grinnell, A. D., and H-U. Schnitzler. 1977. Directional sensitivity of echolocation in the horseshoe bat, *Rhinolophus ferrumequinum*. II Behavioral directionality of hearing. *J. Comp. Physiol.*, 116:63–76.

Grummon, R. A., and A. Novick. 1963. Obstacle avoidance in the bat *Macrotus mexicanus*. *Physiol. Zool.*, 36:361–369.

Gupta, B. B. 1967. The histology and musculature of the plagiopatagium in bats. *Mammalia*, 31:313–321.

Gustafson, A. W., and B. J. Weir (eds.). 1979. *Comparative Aspects of Reproduction in Chiroptera*. J. Reprod. Fert., Symp. Report 14.

Gustafson, V., and H-U. Schnitzler. 1979. Echolocation and obstacle avoidance in the hipposiderid bat, *Asellia tridens*. *J. Comp. Physiol.*, 131:161–168.

Guthrie, M. J. 1933. Notes on the seasonal movements and habits of some cave bats. *J. Mamm.*, 14:1–19.

Habersetzer, J. 1981. Adaptive echolocation sounds in the bat *Rhinopoma hardwickei*, a field study. *J. Comp. Physiol.*, 144:559–566.

Hall, E. R. 1981. *The Mammals of North America*, second edition. New York: John Wiley and Sons.

Haraiwa, Y. K., and T. Uchida. 1956. Fertilization capacity of spermatozoa stored in the uterus after copulation in the fall. *Sci. Bull. Fac. Agric. Kyushu Univ.*, 31:565–574.

Harris, A. H. 1974. *Myotis yumanensis* in interior southwestern North America, with comments on *Myotis lucifugus*. *J. Mamm.*, 55:589–607.

Harrison, T. M. 1983. Patterns of feeding and habitat use by little brown bats *Myotis lucifugus* (Chiroptera: Vespertilionidae) over, and around Lake Opinicon, Ontario, Canada. M.Sc. Thesis, Department of Biology, Carleton University, Ottawa, Canada.

Hayward, B. J. 1970. The natural history of the cave bat *Myotis velifer*. *West. N.M. Univ. Res. Sci.*, 1:1–74.

Heinrich, B. 1979. *Bumblebee Economics.* Cambridge: Harvard Univ. Press.

Heithaus, E. R. 1982. Coevolution between bats and plants. In *Ecology of Bats*, ed. T. H. Kunz, pp. 327–368. New York: Plenum Press.

Heithaus, E. R., and T. H. Fleming. 1978. Foraging movements of a frugivorous bat, *Carollia perspicillata* (Phyllostomatidae). *Ecol. Monog.*, 48:127–143.

Hendrichs, H. 1983. On the evolution of social structure in mammals. In *Advances in the Study of Mammalian Behavior*, ed. J. F. Eisenberg and D. G. Kleiman, pp. 738–750. Am. Soc. Mamm. Spec. Pub. no. 7.

Henson, O. W., Jr. 1970a. The ear and audition. In *Biology of Bats*, vol. 2, ed. W. A. Wimsatt, pp. 181–264. New York: Academic Press.

———1970b. The central nervous system. In *Biology of Bats*, vol. 2, ed. W. A. Wimsatt, pp. 58–152. New York: Academic Press.

Herd, R. M., and M. B. Fenton. 1983. An electrophoretic, morphological, and ecological investigation of a putative hybrid zone between *Myotis lucifigus* and *Myotis yumanensis*. *Can. J. Zool.*, 61:2029–2050.

Hermanson, J. W., and J. S. Altenbach. 1983. The functional anatomy of the shoulder of the pallid bat, *Antrozous pallidus*. *J. Mamm.*, 64:62–75.

Hill, J. E., and J. D. Smith. 1984. *Bats, a Natural History.* London: British Museum (Natural History).

Holroyd, G. L. 1983. Foraging strategies and food of a swallow guild. Ph.D. Thesis, Department of Zoology, University of Toronto, Toronto, Canada.

Hope, G. M., and K. P. Bhatnagar. 1979a. Electrical response of microchiropteran species. *Experientia*, 35:1189–1191.

———1979b. Effect of light adaptation on electrical responses of the retinas of four species of bats. *Experientia*, 35:1191–1193.

Howell, D. J. 1974. Acoustic behavior and feeding in glossophagine bats. *J. Mamm.*, 55:293–308.

———1979. Flock foraging in nectar feeding bats, *Leptonycteris sanborni*, advantages to the bats and to host plants. *Am. Nat.*, 114(1):23–50.

———1980. Adaptive variation in diets of desert bats has implications for evolution of feeding strategies. *J. Mamm.*, 61:730–733.

Howell, D. J., and D. Burch. 1974. Food habits of some Costa Rican bats. *Rev. Biol. Trop.*, 21:281–294.

Howell, K. M. 1980. *Triaenops persicus afer* (Hipposideridae) and conditions of anoxia. *Bat Research News*, 21:26–30.

Humphries, D. A., and P. M. Driver. 1970. Protean defence by prey animals. *Oecologia*, 5:285–302.

Jen, P. H-S., and N. Suga. 1976. Coordinated activities of middle-ear and laryngeal muscles in echolocating bats. *Science*, 191:950–952.

Jepsen, G. L. 1970. Bat origins and evolution. In *Biology of Bats*, vol. I, ed. W. A. Wimsatt, pp. 1–64. New York: Academic Press.

Jones, C. 1967. Growth, development and wing loading in the evening bat, *Nycticeius humeralis*. *J. Mamm.*, 48:1–19.

Jones, C., and R. D. Suttkus. 1975. Notes on the natural history of *Plecotus rafinesquii*. *Occ. Pap. Mus. Zool. La. State Univ.*, 47:1–14.

Jones, J. K., Jr., and H. H. Genoways. 1970. Chiropteran systematics. In *About Bats*, ed. B. H. Slaughter and D. W. Walton, pp. 3–21. Dallas: Southern Methodist Univ. Press.

Kallen, F. C. 1977. The cardiovascular systems of bats: structure and function. In *Biology of Bats*, vol. III, ed. W. A. Wimsatt, pp. 290–518. New York: Academic Press.

Kallen, F. C., and C. Gans. 1972. Mastication in the little brown bat, *Myotis lucifugus*. *J. Morph.*, 136:385–420.

Kamper, R., and U. Schmidt. 1977. Die Morphologie der Nasenhole bei einigen neotropischen Chiropteren. *Zoomorphologie*, 87:3–19.

Kick, S. A. 1982. Target-detection by the echolocating bat, *Eptesicus fuscus*. *J. Comp. Physiol.*, 145:431–435.

Kingdon, J. 1974. *East African Mammals, an Atlas of Evolution in Africa*, vol. IIA. New York: Academic Press.

Kitchener, D. J. 1975. Reproduction in female Gould's wattled bat *Chalinolobus gouldii* (Gray) (Vespertilionidae) in western Australia. *Aust. J. Zool.*, 23:29–42.

Kitchener, D. J., and S. A. Halse. 1978. Reproduction in female *Eptesicus regulus* (Thomas) (Vespertilionidae) in south-western Australia. *Aust. J. Zool.*, 26:257–267.

Kleiman, D. G. 1969. Maternal care, growth rate, and development in the noctule *(Nyctalus noctula)*, pipistrelle *(Pipistrellus pipstrellus)*, and serotine *(Eptesicus serotinus)* bats. *J. Zool. Lond.*, 157:187–211.

Kleiman, D. G., and T. M. Davis. 1979. Ontogeny and maternal care. In *Biology of Bats of the New World Family Phyllostomatidae*, part III, ed. R. J. Baker, J. K. Jones, and D. C. Carter, pp. 387–402. Lubbock: Texas Tech Univ. Press.

Kleiman, D. G., and P. A. Racey. 1969. Observations of noctule bats *(Nyctalus noctula)* breeding in captivity. *Lynx*, 10:65–77.

Kolmer, W. 1924. Über die augen der fledermause. *Z. Ges. Anat.*, 73:645–658.

Komisaruk, B. R. 1977. The role of rhythmical brain activity in sensory-motor integration. In *Progress in Psychobiology and Physiological Psychology*, ed. J. M. Sprague and A. N. Epstein, pp. 55–90. New York: Academic Press.

Koopman, K. F., and J. K. Jones, Jr. 1970. Classification of bats. In *About Bats*, ed. B. H. Slaughter and D. W. Walton, pp. 22–28. Dallas: Southern Methodist Univ. Press.

Koopman, K. F., and G. T. MacIntyre. 1980. Phylogenetic analysis of chiropteran dentition. In *Proceedings Fifth International Bat Research Conference*, ed. D. E. Wilson and A. L. Gardner, pp. 279–288. Lubbock: Texas Tech Univ. Press.

Kratky, J. 1970. Postnatale entwicklung des grossmausohrs, *Myotis myotis* (Borkhausen, 1797). *Acta Soc. Zool. Bohem.*, 33:202–218.

Krishna, A. 1978. Aspects of reproduction in some Indian bats. Ph.D dissertation. Banaras Hindu Univ., Varanasi.

Krishna, A., and C. J. Dominic. 1982. Differential rates of fetal growth in two successive pregnancies in the emballonurid bat, *Taphozous longimanus* Hardwicke. *Biol. Reprod.*

Kulzer, E. 1958. Untersuchungen über die Biologie von Flughunden der Gattung *Rousettus*. *Z. Morphol. Oekol. Tiere.*, 47:374–402.

———1961. Über die Biologie der Nil-Flughunde *(Rousettus aegyptiacus). Natur. Volk,* 91:219–228.

———1962. Über die Jugendentwicklung der Angola-Bulldogfledermaus *Tadarida (Mops) condylura* (A. Smith, 1933) (Molossidae). *Saugtierk. Mitt.,* 10:116–124.

Kunz, T. H. 1973. Population studies of the cave bat, *Myotis velifer:* reproduction, growth and development. *Occ. Pap. Mus. Nat. Hist. Univ. Kansas,* 15:1–43.

———1974. Reproduction, growth, and mortality of the vespertilionid bat, *Eptesicus fuscus* in Kansas. *J. Mamm.,* 55:1–13.

———1982a. Roosting ecology. In *Ecology of Bats,* ed. T. H. Kunz, pp. 1–56. New York: Plenum Press.

———(ed.) 1982b. *Ecology of Bats.* New York: Plenum Publishing Corp.

———(ed.) in press. Ecological and behavioral methods for the study of bats. Washington: Smithsonian Institution Press.

Kunz, T. H., and E. L. P. Anthony. 1982. Age estimation and postnatal growth rates in the bat *Myotis lucifugus. J. Mamm.,* 63:23–32.

Kunz, T. H., P. V. August, and C. D. Burnett. 1983. Harem social organization in cave roosting *Artibeus jamaicensis* (Chiroptera: Phyllostomidae). *Biotropica,* 15:133–138.

Kürten, L., and U. Schmidt. 1982. Thermoperception in the common vampire bat *(Desmodus rotundus). J. Comp. Physiol.,* 146:223–228.

Lawrence, B., and A. Novick. 1963. Behavior as a taxonomic clue: relationships of *Lissonycteris* (Chiroptera). *Brevoria,* 184:1–16.

Lawrence, B. D., and J. A. Simmons. 1982a. Echolocation in bats: the external ear and perception of the vertical positions of targets. *Science,* 218:481–483.

———1982b. Measurements of atmospheric attenuation at ultrasonic frequencies and the significance for echolocation by bats. *J. Acoust. Soc. Am.,* 71:585–590.

Leen, N., and A. Novick. 1969. *The World of Bats.* New York: Holt, Rinehart and Winston.

Leonard, M. L., and M. B. Fenton. 1983. Habitat use by spotted bats (*Euderma maculatum,* Chiroptera: Vespertilionidae): roosting and foraging behaviour. *Can. J. Zool.,* 61:1487–1491.

———1984. Echolocation calls of *Euderma maculatum* (Chiroptera: Vespertilionidae): use in orientation and communication. *J. Mamm.,* 65:122–126.

Licht, P., and P. Leitner. 1967. Behavioral responses to high temperatures in three species of California bats. *J. Mamm.,* 48:52–61.

Likhachev, G. N. 1961. Use by bats of bird nesting boxes. *Prioksho-Terranisk Gos. Zapovednika,* 3:85–156.

Long, G. R., and H-U. Schnitzler. 1975. Behavioural audiograms for the bat, *Rhinolophus ferrumequinum. J. Comp. Physiol.,* 100:211–219.

Luckett, W. P. 1980. The use of fetal membrane data in assessing chiropteran membrane phylogeny. In *Proceedings Fifth International Bat Research Conference,* ed. D. E. Wilson and A. L. Gardner, pp. 245–266. Lubbock: Texas Tech Univ. Press.

McCracken, G. F., and J. W. Bradbury. 1977. Paternity and genetic heterogeneity in the polygynous bat, *Phyllostomus hastatus. Science,* 198:303–306.

———1981. Social organization and kinship in the polygynous bat, *Phyllostomus hastatus. Behav. Ecol. Sociobiol.,* 8:11–34.

———1984. Communal nursing in Mexican free-tailed bat maternity colonies. *Science,* 223:1090–1091.

McNab, B. K. 1973. Energetics and the distribution of vampires. *J. Mamm.,* 54:131–144.

———1982. Evolutionary alternatives in the physiological ecology of bats. In *Ecology of Bats,* ed. T. H. Kunz, pp. 151–200. New York: Plenum Press.

McWilliam, A. N. 1982. Adaptive responses to seasonality in four species of Microchiroptera in coastal Kenya. Ph.D. Thesis, University of Aberdeen, Aberdeen, Scotland.

Madhaven, A., and A. Gopalakrishna. 1978. Breeding habits and associated phenomena in some Indian bats. IV. *Hipposideros fulvus fulvus* (Gray)-Hipposideridae. *J. Bombay Nat. Hist. Soc.,* 75:96–103.

Maeda, K. 1972. Growth and development of the large noctule, *Nyctalus lasiopterus,* Schreber. *Mammalia,* 36:269–278.

———1974. Eco-ethologie de la grande noctule, *Nyctalus lasiopterus,* a Sapporo, Japan. *Mammalia,* 38:461–487.

———1976. Growth and development of the Japanese large-footed bat, *Myotis macrodactylus.* I External characters and breeding habits. *J. Growth,* 15:29–40.

Mainoya, J. R., and K. M. Howell. 1977. Histology of the frontal sac in three species of leaf-nosed bats (Hipposideridae). *E. Afr. Wildl. J.,* 15:147–155.

Marler, P. 1983. Monkey calls: how they are perceived and what do they mean? In *Advances in the Study of Mammalian Behavior,* ed. J. F. Eisenberg and D. G. Kleiman, pp. 343–356. Am. Soc. Mamm. Spec. Pub. no. 7.

Marshall, A. G. 1981. *Ecology of Ectoparasitic Insects.* London: Academic Press.

———1982. Ecology of insects ectoparasitic on bats. In *Ecology of Bats,* ed. T. H. Kunz, pp. 369–402. New York: Plenum Press.

Marshall, A. G., and A. N. McWilliam 1982. Ecological observations on epomophorine fruit-bats (Megachiroptera) in west African savanna woodland. *J. Zool. Lond.,* 198:53–67.

Masterson, F. A., and S. R. Ellins. 1974. The role of vision in the orientation of the echolocating bat. *Behavior,* 51:88–98.

Matsumura, S. 1979. Mother-infant communication in a horseshoe bat *(Rhinolophus ferrumequinum nippon):* development of vocalizations. *J. Mamm.,* 60:76–84.

———1981. Mother-infant communication in a horseshoe bat *(Rhinolophus ferrumequinum nippon):* vocal communication in three-week-old infants. *J. Mamm.,* 62:20–28.

Matthews, L. H. 1941. Notes on the genitalia and reproduction of some African bats. *Proc. Zool. Soc. London,* 111:289–346.

Medway, Lord. 1971. Observations of social and reproductive biology of the bent-winged bat *Miniopterus australis* in northern Borneo. *J. Zool. Lond.,* 165:261–273.

———1972. Reproductive cycles of the flat-headed bats *Tylonycteris pachypus* and *T. robustula* (Chiroptera: Vespertilionidae) in a humid equatorial environment. *J. Linn. Soc. Lond. Zool.,* 159:33–61.

Menaker, M. 1962. Hibernation—hypethermia: an annual cycle of response to low temperature in the bat *Myotis lucifugus. J. Cell. Comp. Physiol.*, 57:81–86.

Miller, G. S., Jr. 1907. Families and genera of bats. *U.S. Nat. Mus. Bull.* 57:1–282.

Miller, L. A. 1982. The orientation and evasive behavior of insects to bat cries. In *Exogenous and Endogenous Influences on Metabolic and Neural Control*, ed. A. D. F. Addink and N. Spronk, pp. 393–405. Oxford: Pergamon Press.

Miller, L. A., and H. J. Degn. 1981. The acoustic behavior of four vespertilionid bats studied in the field. *J. Comp. Physiol.*, 142:67–74.

Miller, R. E. 1939. The reproductive cycle in male bats of the species *Myotis lucifugus* and *Myotis grisescens. J. Morphol.*, 64:267–295.

Möhres, F. P. 1967. Communicative characters of sonar signals in bats. In *Animal Sonar Systems*, vol. 2, ed. R-G. Busnel, pp. 939–945. NATO Advanced Study Institute.

Morrison, D. W. 1978. Lunar phobia in a neotropical fruit bat, *Artibeus jamaicensis* (Chiroptera: Phyllostomatidae). *Anim. Behav.*, 26:852–855.

———1979. Apparent male defense of tree hollows in the fruit bat, *Artibeus jamaicensis. J. Mamm.*, 60:11–15.

———1980. Foraging and day-roosting dynamics of canopy fruit bats in Panama. *J. Mamm.*, 61:20–29.

Morton, E. S. 1977. On the occurrence and significance of motivation-structural rules in some bird and mammal sounds. *Am. Nat.*, 111:855–867.

Mueller, H. C., and N. S. Mueller. 1979. Sensory basis for spatial memory in bats. *J. Mamm.*, 60:198–201.

Muller-Schwarze, D. 1983. Scent glands in mammals and their function. In *Advances in the Study of Mammalian Behavior*, ed. J. F. Eisenberg and D. G. Kleiman, pp. 150–197. Am. Soc. Mamm. Spec. Pub. no. 7.

Murphy, C. J., H. C. Howland, G. G. Kwiecinski, T. Kern, and F. Kallen. 1983. Visual accommodation in the flying fox *(Pteropus giganteus)*. Vision Res., 23:617–620.

Mutere, F. A. 1967. The breeding biology of equatorial vertebrates: reproduction in the fruit bat *Eidolon helvum*, at latitude 0°20'N. *J. Zool. Lond.*, 153:153–161.

Myers, P. 1977. Patterns of reproduction in four species of vespertilionid bats in Paraguay. *Univ. Calif. Pub. Zool.*, 107:1–41.

Nellis, D. W., and C. F. Ehle. 1977. Observations on the behavior of *Brachyphylla cavernarum* (Chiroptera). *Mammalia*, 41:403–409.

Nelson, J. E. 1964. Vocal communication in Australian flying foxes (Pteropodidae: Megachiroptera). *Z. Tierpsychol.*, 21:857–870.

———1965. Behaviour of Australian Pteropodidae (Megachiroptera). *Anim. Behav.*, 13:544–557.

Neuweiler, G. 1967. Interaction of other sensory systems with the sonar system. In *Animal Sonar Systems*, vol. 1, ed. R-G. Busnel, pp. 509–534. NATO Advanced Study Institute.

———1969. Verhaltensbeobachtungen an einer Indischen Flughundkolonie *(Pteropus g. giganteus). Z. Tierpsychol.*, 26:166–199.

———1980a. Auditory processing of echoes: peripheral processing. In *Animal Sonar Systems*, ed. R-G. Busnel and J. F. Fish, pp. 519–548. NATO Advanced Study Institutes Series A28. New York: Plenum Press.

———1980b. How bats detect flying insects. *Physics Today*, 33(8):34–40.

Norberg, U. M. 1969. An arrangement giving a stiff leading edge to the hand wing in bats. *J. Mamm.*, 50:766–770.

———1976a. Some advanced flight manoeuvres of bats. *J. Exp. Biol.*, 64:489–495.

———1976b. Aerodynamics, kinematics and energetics of horizontal flapping flight in the long-eared bat, *Plecotus auritus. J. Exp. Biol.*, 65:179–212.

———1976c. Aerodynamics of hovering flight in the long-eared bat, *Plecotus auritus. J. Exp. Biol.*, 65:459–470.

Novacek, M. J. 1980. Phylogenetic analysis of the chiropteran auditory region. In *Proceedings Fifth International Bat Research Conference*, ed. D. E. Wilson and A. L. Gardner, pp. 317–330. Lubbock: Texas Tech Univ. Press.

Novick, A. 1958. Orientation in paleotropical bats, I Microchiroptera. *J. Exp. Zool.*, 138:81–154.

———1977. Acoustic orientation. In *Biology of Bats*, vol. III, ed. W. A. Wimsatt, pp. 289. New York: Academic Press. Novick, A., and B. A. Dale. 1971. Foraging behavior in fishing bats and their insectivorous relatives. *J. Mamm.*, 52:817–818.

O'Farrell, M. J., and E. H. Studier. 1973. Reproduction, growth and development in *Myotis thysanodes* and *Myotis lucifugus. Ecology*, 54:18–30.

Orr, R. T. 1954. Natural history of the pallid bat, *Antrozous pallidus* (LeConte). *Proc. California Acad. Sci.*, 28:165–246.

———1970. Development: prenatal and postnatal. In *Biology of Bats*, vol. I, ed. W. A. Wimsatt, pp. 217–232. New York: Academic Press.

O'Shea, T. J. 1980. Roosting, social organization and the annual cycle in a Kenya population of the bat, *Pipistrellus nanus. Z. Tierpsychol.*, 53:171–195.

Pagels, C. F., and C. Jones. 1974. Growth and development of the free-tailed bat, *Tadarida brasiliensis cynocephalus* (LeConte). *Southwest. Nat.*, 19:267–276.

Parkinson, A. 1979. Morphologic variation and hybridization in *Myotis yumanensis sociabilis* and *Myotis lucifugus carissima. J. Mamm.*, 60:489–504.

Pearson, O. P., M. R. Koford, and A. K. Pearson. 1952. Reproduction of the lump-nosed bat *(Corynorhinus rafinesqui)* in California. *J. Mamm.*, 33:273–320.

Pedler, C., and R. Tilley. 1969. The retina of a fruit bat. *Vision Research*, 9:909–922.

Peterson, R. L. 1965a. A review of the bats of the genus *Ametrida*, family Phyllostomidae. *Life Sci. Contr., R. Ont. Mus.*, 65:1–13.

———1965b. A review of the flat-headed bats of the family Molossidae from South America and Africa. *Life Sci. Contr., R. Ont. Mus.*, 64:1–32.

Peyre, A., and M. Herlant. 1967. Ovo-implantation différée et déterminisme hormonal chez Minioptère *Miniopterus schreibersi* K. (Chiroptère). *C. R. Seances Soc. Biol. Paris*, 161:1779–1782.

Phillips, C. J. 1971. The dentition of the glossophagine bats; development, morphological characteristics, variation, pathology and evolution. Univ. Kansas Mus. Nat. Hist. Misc. Pub., no. 54:1–138.

Pierson, E. D., V. M. Sarich, J. M. Lowenstein, and M. J. Daniel. 1982. *Mystacina* is a phyllostomatoid bat. *Bat Research News*, 23:78.

Porter, F. L. 1978. Roosting patterns and social behavior in captive *Carollia perspicillata*. *J. Mamm.*, 59:627–630.

———1979a. Social behavior in the leaf-nosed bat, *Carollia perspicillata*, I social organization. *Z. Tierpsychol.*, 49:406–417.

———1979b. Social behavior in the leaf-nosed bat, *Carollia perspicillata*. II social communication. *Z. Tierpsychol.*, 50:1–8.

Porter, F. L., and G. F. McCracken. 1983. Social behavior and allozyme variation in a captive colony of *Carollia perspicillata*. *J. Mamm.*, 64:295–298.

Poussin, C., and J. A. Simmons. 1982. Low frequency hearing sensitivity in the echolocating bat, *Eptesicus fuscus*. *J. Acoust. Soc. Am.*, 72:340–342.

Pye, J. D. 1978. Some preliminary observations on flexible echolocation systems. In *Proceedings Fourth International Bat Research Conference*, ed. R. J. Olembo, J. B. Castelino, and F. A. Mutere, pp. 127–136. Nairobi: Kenya Lit. Bureau.

———1980. Echolocation signals and echoes in air. In *Animal Sonar Systems*, ed. R-G. Busnel and J. F. Fish, pp. 309–354. NATO Advanced Study Institutes A28. New York: Plenum Press.

Quay, W. B. 1970. Integument and derivatives. In *Biology of Bats*, ed. W. A. Wimsatt, pp. 1–56. New York: Academic Press.

Racey, P. A. 1973. The viability of spermatozoa after prolonged storage by male and female European bats. *Period. Biol.*, 75:201–205.

———1974. Aging and assessment of reproductive status of pipistrelle bats, *Pipistrellus pipistrellus*. *J. Zool. Lond.*, 173:264–271.

———1975. The prolonged survival of spermatozoa in bats. In *The Biology of the Male Gamete*, ed. J. G. Duckett and P. A. Racey, pp. 385–416. London: Academic Press.

———1979. The prolonged storage and survival of spermatozoa in Chiroptera. *J. Reprod. Fert.*, 56:391–402.

———1982. Ecology of bat reproduction. In *Ecology of Bats*, ed. T. H. Kunz, pp. 57–104. New York: Plenum Press.

Rakhmatulina, I. K. 1972. The breeding, growth and development of pipistrelles in Azerbaidzhan. *Sov. J. Ecol.*, 2:131–136.

Ramakrishna, P. A. 1951. Studies on the reproduction in bats. I. Some aspects of the reproduction in the oriental vampires, *Lyroderma lyra lyra* (Geoffroy) and *Megaderma spasma* (Linn.). *J. Mysore Univ.*, 12:107–118.

Ramakrishna, P. A., and K. V. B. Rao. 1977. Reproductive adaptations in the Indian rhinolophid bat. *Rhinolophus rouxi* (Temminck). *Curr. Sci.*, 46:270–271.

Ransome, R. 1980. *The Greater Horseshoe Bat*. Poole: Blanford Press.

Rasweiler, J. J. Iv. 1977. The care and management of bats as laboratory animals. In *Biology of Bats*, vol. III, ed. W. A. Wimsatt, pp. 519–617. New York: Academic Press.

Rice, D. W. 1957. Life history and ecology of *Myotis austroriparius* in Florida. *J. Mamm.*, 38:15–32.

Richardson, E. G. 1977. The biology and evolution of the reproductive cycle of *Miniopterus schreibersii* and *M. australis* (Chiroptera: Vespertilionidae). *J. Zool. Lond.*, 183:353–375.

Roberts, L. H. 1972. Variable resonance in constant frequency bats. *J. Zool. Lond.*, 166:337–348.

———1975. Confirmation of the echolocation pulse production mechanism of *Rousettus*. *J. Mamm.*, 56:218–220.

Roeder, K. D. 1967. *Nerve Cells and Insect Behavior*, revised edition. Cambridge: Harvard Univ. Press.

Rosevear, D. R. 1965. *The Bats of West Africa*. London: Trustees of the British Museum (Natural History).

Roth, C. E. 1957. Notes on maternal care in *Myotis lucifugus*. *J. Mamm.*, 38:122–123.

Ryan, M. J., M. D. Tuttle, and R. M. R. Barclay. 1983. Behavioral responses of the frog-eating bat, *Trachops cirrhosus*, to sonic frequencies. *J. Comp. Physiol.*, 150:413–418.

Rybaur, P. 1971. On the problems of practical use of the ossification of bones as age criteria in the bats (Microchiroptera). *Pr. Studie-Purir. Pardubice*, 3:97–121.

Schmidt, U. 1972. Social calls of juvenile vampire bats *(Desmodus rotundus)* and their mothers. *Bonn. Zool. Beitr.*, 4:310–316.

———1978. *Vampirfledermause*. Wittenberg Lutherstadt: Die Neue Brehmn-Bucherei. A. Ziemsen Verlag.

Schmidt, U., and U. Manske. 1973. Die Jugendentwicklung der Vampir-fledermause *(Desmodus rotundus)*. *Z. Saugetierk*, 38:14–33.

Schneider, R. 1957. Morphologische untersuchungen am Gehirn der Chiroptera (Mammalia). *Abh. Senckb. Naturforsch. Ges.*, 495:1–92.

Schnitzler, H-U., and E. Flieger. 1983. Detection of oscillating target movements by echolocation in the greater horseshoe bat. *J. Comp. Physiol.*, 153:385–391.

Schnitzler, H-U., and A. D. Grinnell. 1977. Directional sensitivity of echolocation in the horseshoe bat, *Rhinolophus ferrumequinum* I. Directionality of sound emission. *J. Comp. Physiol.*, 116:51–61.

Schuller, G., and G. Pollack. 1979. Disproportionate frequency representation in the inferior colliculus of Doppler-compensating greater horseshoe bats: evidence for an acoustic fovea. *J. Comp. Physiol.*, 132:47–54.

Sherman, H. B. 1937. Breeding habits of the free-tailed bat. *J. Mamm.*, 18:176–187.

Short, H. L. 1961. Growth and development of Mexican free-tailed bats. *Southwest. Nat.*, 6:156–163.

Sigé, B. 1977. Les insectivores et chiroptères du Paléogène moyen d'Europe dans l'histoire des faunes des mammifères sur ce continent. *J. Paleont. Soc. India*, 20:178–190.

Sigé, B., and D. E. Russell. 1980. Complements sur les chiroptères de l'Eocene moyen d'Europe. Les genres *Palaeochiropteryx* et *Cecilionycteris*. *Palaeovertebrata, Mem. J. R. Iavovat:*91–126.

Simmons, J. A. 1971. The sonar receiver of the bat. *Ann. N. Y. Acad. Sci.*, 188:161–174.

Simmons, J. A., M. B. Fenton, and M. J. O'Farrell. 1979. Echolocation and pursuit of prey by bats. *Science*, 203:16–21.

Simmons, J. A., M. B. Fenton, W. R. Ferguson, M. Jutting, and J. Palin. 1979.

Apparatus for research on animal ultrasonic signals. *Life Sci. Misc. Pub., R. Ont. Mus.,* 1–31.

Simmons, J. A., W. A. Lavender, B. A. Lavender, J. E. Childs, K. Hulebak, M. R. Rigden, J. Sherman, B. Woolman, and M. J. O'Farrell. 1978. Echolocation by free-tailed bats *(Tadarida). J. Comp. Physiol.,* 125:191–199.

Simmons, J. A., and R. A. Stein. 1980. Acoustic imaging in bat sonar: echolocation signals and the evolution of echolocation. *J. Comp. Physiol.,* 135:61–84.

Simpson, G. G. 1980. *Spendid Isolation, the Curious History of South American Mammals.* New Haven: Yale Univ. Press.

Slaughter, B. H. 1970. Evolutionary trends of chiropteran dentition. In *About Bats,* ed. B. H. Slaughter and D. W. Wilson, pp. 51–83. Dallas: Southern Methodist Univ. Press.

Slaughter, B. H., and D. W. Walton (eds.). 1970. *About Bats, a Chiropteran Symposium.* Dallas: Southern Methodist Univ. Press.

Sluiter, J. W. 1954. Sexual maturity in bats of the genus *Myotis* II. Females of *M. mystacinus* and *M. emarginatus. Proc. K. Ned. Akad. Wet.,* 57:696–700.

———1960. Reproductive rate of the bat *Rhinolophus hipposideros. Proc. K. Ned. Akad. Wet.,* 63:383–393.

———1961. Sexual maturity in males of the bat *Myotis myotis. Proc. K. Ned. Akad. Wet.,* 64:243–249.

Sluiter, J. W., and M. Bouman. 1951. Sexual maturity in bats of the genus *Myotis.* I Size and histology of reproductive organs during hibernation in connection with age and wear of the teeth in female *Myotis myotis* and *Myotis emarginatus. Proc. K. Ned. Akad. Wet.,* 54:595–601.

Sluiter, J. W., and P. F. van Heerdt. 1966. Seasonal habits of the noctule bat *(Nyctalus noctula). Arch. Nerrl. Zool.,* 16:423–439.

Smith, J. D. 1972. Systematics of the chiropteran family Mormoopidae. *Univ. Kansas, Mus. Nat. Hist. Misc. Pub.,* 56:1–132.

———1976. Chiropteran evolution. In *Biology of Bats of the New World Family Phyllostomatidae,* part 1, ed. R. J. Baker, J. K. Jones, Jr., and D. C. Carter, pp. 49–70. Lubbock: Special Publications, The Museum, Texas Tech Univ.

———1977. Comments on flight and the evolution of bats. In *Major Patterns in Vertebrate Evolution,* ed. M. K. Hecht, P. C. Goody, and B. M. Hecht, pp. 427–437. New York: NATO Advanced Study Institutes, A14, Plenum Press.

———1980. Chiroptera phylogenetics: introduction. In *Proceedings Fifth International Bat Research Conference,* ed. D. E. Wilson and A. L. Gardner, pp. 233–244. Lubbock: Texas Tech Univ. Press.

Smith, J. D., and G. Madkour. 1980. Penial morphology and the question of chiropteran phylogeny. In *Proceedings Fifth International Bat Research Conference,* ed. D. E. Wilson and A. L. Gardner, pp. 347–366. Lubbock: Texas Tech Univ. Press.

Spencer-Booth, Y. 1970. The relationships between mammalian young and conspecifics other than mothers and peers: a review. In *Advances in Animal Behavior,* vol. 3, ed. D. S. Lehrman, R. A. Hinde, and E. Shaw, pp. 119–194. New York: Academic Press.

Stebbings, R. E. 1966. A population study of bats of the genus *Plecotus. J. Zool. Lond.,* 150:53–75.

Stoddart, D. M. 1980. *The Ecology of Vertebrate Olfaction.* London: Chapman and Hall.

Strickler, T. L. 1978. *Functional Osteology and Myology of the Shoulder Girdle in the Chiroptera.* Basel: S. Karger.

Studier, E. H. 1969. Respiratory ammonia filtration, mucous composition and ammonia tolerance in bats. *J. Exp. Zool.,* 170:253–258.

Suthers, R. A. 1965. Acoustic orientation by fish-catching bats. *J. Exp. Zool.,* 158:319–348.

———1967. Comparative echolocation by fishing bats. *J. Mamm.,* 48:79–87.

———1970a. Vision, olfaction, taste. In *Biology of Bats,* vol. II, ed. W. A. Wimsatt, pp. 265–309. New York: Academic Press.

———1970b. A comment on the role of the choroidal papillae in the fruit bat retina. *Vision Research,* 10:921–922.

Suthers, R. A., and M. R. Braford, Jr. 1980. Visual systems and the evolutionary relationships of the Chiroptera. In *Proceedings Fifth International Bat Research Conference,* ed. D. E. Wilson and A. L. Gardner, pp. 331–346. Lubbock: Texas Tech Univ. Press.

Suthers, R. A., and J. M. Fattu. 1982. Selective laryngeal neurotomy and the control of phonation by the echolocating bat, *Eptesicus. J. Comp. Physiol.,* 145:529–537.

Suthers, R. A., and D. H. Hector. 1982. Mechanism for production of echolocating clicks by the grey swiftlet, *Collocalia spodiopygia. J. Comp. Physiol.,* 148:457–470.

Suthers, R. A., and C. A. Summers. 1980. Behavioral audiogram and masked thresholds of the megachiropteran echolocating bat, *Rousettus. J. Comp. Physiol.,* 136:277–283.

Suthers, R. A., and N. E. Wallis. 1970. Optics of the eyes of echolocating bats. *Vision Research,* 10:1165–1173.

Swift, S. M. 1981. Foraging, colonial and maternal behaviour of bats in northeast Scotland. Ph.D Thesis, University of Aberdeen, Aberdeen, Scotland.

Thomas, D. W. 1979. Plans for a lightweight, inexpensive radio transmitter. In *A Handbook on Biotelemetry and Radio Tracking,* ed. C. J. Amlaner, Jr., and D. W. MacDonald, pp. 175–179. New York: Pergamon Press.

———1982. The ecology of an African savanna fruit bat community: resource partitioning and role in seed dispersal. Ph.D Thesis, University of Aberdeen, Aberdeen, Scotland.

———1983. The annual migrations of three species of west African fruit bats (Chiroptera: Pteropodidae). *Can. J. Zool.,* 61:2266–2272.

Thomas, D. W., and M. B. Fenton. 1978. Notes on the dry season roosting and foraging behavior of *Epomophorus gambianus* and *Rousettus aegyptiacus* (Chiroptera: Pteropodidae). *J. Zool. Lond.,* 186:403–406.

Thomas, D. W., M. B. Fenton, and R. M. R. Barclay. 1979. Social behavior of the little brown bat, *Myotis lucifugus.* 1. mating behavior. *Behav. Ecol. Sociobiol.,* 6:129–136.

Thomas, S. P. 1975. Metabolism during flight in two species of bats, *Phyllostomus hastatus* and *Pteropus gouldii. J. Exp. Biol.,* 63:273–293.

Thompson, D., and M. B. Fenton. 1982. Echolocation and feeding behavior of *Myotis adversus* (Chiroptera: Vespertilionidae). *Aust. J. Zool.,* 30:543–546.

Thomson, C. E. 1980. Mother-infant interactions in free-living little brown bats, *Myotis lucifugus* (Chiroptera: Vespertilionidae). M.Sc. Thesis, Department of Biology, Carleton University, Ottawa, Canada.

Timm, R. M., and J. Mortimer. 1976. Selection of roost sites by Honduran white bats, *Ectophylla alba* (Chiroptera; Phyllostomatidae). *Ecology*, 57:385–389.

Trune, D. R., and C. N. Slobodchikoff. 1976. Social effects of roosting on the metabolism of the pallid bat *(Antrozous pallidus)*. *J. Mamm.*, 57:656–663.

Turner, D. C. 1975. *The Vampire Bat, a Field Study in Behavior and Ecology.* Baltimore: Johns Hopkins University Press.

Turner, D. C., A Shaughnessy, and E. Gould. 1972. Individual recognition between mother and infant bats *(Myotis).* In *Animal Orientation and Navigation,* ed. S. R. Gallerk, K. Schmidt-Koenig, G. J. Jacobs, and R. E. Belleville, pp. 365–371. NASA Sp-262.

Turner, G. E. 1970. Sexual dimorphism in eleven species of New World and eighteen species of Old World bats of the family Molossidae. M.Sc. Thesis, University of Toronto, Toronto, Canada.

Tuttle, M. D. 1974. An improved trap for bats. *J. Mamm.*, 55:475–477.

———1975. Population ecology of the gray bat *(Myotis grisescens):* factors influencing early growth and development. *Occ. Pap. Mus. Nat. Hist. Univ. Kansas,* no. 36:1–24.

———1982. The amazing frog-eating bat. *National Geographic*, 161:78–91.

Tuttle, M. D., and S. J. Kern. 1981. Bats and public health. *Contr. Biol. Geol. Milwaukee Pub. Mus.*, no. 48:1–11.

Tuttle, M. D., and M. J. Ryan. 1981. Bat predation and the evolution of frog vocalizations in the neotropics. *Science*, 214:677–678.

Tuttle, M. D., and D. Stevenson. 1982. Growth and survival of bats. In *Ecology of Bats,* ed. T. H. Kunz, pp. 105–150. New York: Plenum Press.

Ubelaker, J. E. 1970. Some observations on ecto- and endoparasites of Chiroptera. In *About Bats,* ed. B. H. Slaughter and D. W. Walton, pp. 247–261. Dallas: Southern Methodist Univ. Press.

van der Merwe, M. 1979. Foetal growth curves and seasonal breeding in the Natal clinging bat *Miniopterus schreibersi natalensis. S. Afr. J. Zool.*, 14:17–21.

Van Deusen, H. M. 1968. Carnivorous habits of *Hypsignathus monstrosus. J. Mamm.*, 40:335–336.

Van Valen, L. 1979. The evolution of bats. *Evolutionary Theory,* 4:103–121.

Vaughan, T. A. 1959. Functional morphology of three bats: *Eumops, Myotis, Macrotus. Publ. Univ. Kansas Mus. Nat. Hist.*, 12:1–153.

———1970a. The skeletal system. In *Biology of Bats,* vol. I, ed. W. A. Wimsatt, pp. 98–139. New York: Academic Press.

———1970b. The muscular system. In *Biology of Bats,* vol. I, ed. W. A. Wimsatt, pp. 140–194. New York: Academic Press.

———1970c. Flight patterns and aerodynamics. In *Biology of Bats,* vol. I, ed. W. A. Wimsatt. pp. 195–216. New York: Academic Press.

———1970d. Adaptations for flight in bats. In *About Bats,* ed. B. H. Slaughter and D. W. Wilson, pp. 127–143. Dallas: Southern Methodist Univ. Press.

———1976. Nocturnal behavior of the African false vampire bat *(Cardioderma cor). J. Mamm.*, 57:227–248.

————1977. Foraging behavior of the giant leaf-nosed bat *(Hipposideros commersoni)*. *E. Afr. Wildl. J.*, 15:237–249.

————1980. Opportunistic feeding by two species of *Myotis*. *J. Mamm.*, 61:118–119.

Vaughan, T. A., and T. J. O'Shea. 1976. Roosting ecology of the pallid bat, *Antrozous pallidus*. *J. Mamm.*, 57:19–42.

Vehrencamp, S. L., F. G. Stiles, and J. W. Bradbury. 1977. Observations on the foraging behavior and avian prey of the neotropical carnivorous bat, *Vampyrum spectrum*. *J. Mamm.*, 58:469–478.

Vogler, B., and G. Neuweiler. 1983. Echolocation in the noctule *(Nyctalus noctula)* and horseshoe bat *(Rhinolophus ferrumequinum)*. *J. Comp. Physiol.*, 152:421–432.

Wallace, G. I. 1978. A histological study of the early stages of pregnancy in the bent-winged bat *(Miniopterus schreibersi)* in north-eastern New South Wales, Australia (30°27'S). *J. Zool. Lond.*, 185:519–537.

Walker, E. P., et al. 1975. *Mammals of the World*. Third edition, volume I. Baltimore.

Wallin, L. 1961. Territorialism on the hunting ground of *Myotis daubentoni*. *Saugetierk. Mitt.*, 9:156–159.

Watkins, L. C., and K. A. Shump, Jr. 1981. Behavior of the evening bat, *Nycticeius humeralis* at a nursery roost. *Am. Midl. Nat.*, 105:258–268.

Werner, T. K. 1981. Responses of nonflying moths to ultrasound: the threat of gleaning bats. *Can J. Zool.*, 59:525–529.

Whitaker, J. O., and H. Black. 1976. Food habits of cave bats from Zambia, Africa. *J. Mamm.*, 57:199–204.

Wickler, W., and U. Seibt. 1976. Field studies on the African fruit bat, *Epomophorus wahlbergi* (Sundevall), with special reference to male calling. *Z. Tierpsychol.*, 40:345–376.

Wickler, W., and D. Uhrig. 1969. Verhalten und ökologische Nische der Gelbflugelfledermaus *Lavia frons*. *Z. Tierpsychol.*, 26:726–736.

Wilkinson, G. S. 1983. Blood sharing by the vampire bat, *Desmodus rotundus:* a case for reciprocity and kin selection. *Bat Research News*, 24.

Williams, C. B. 1940. An analysis of four years capture of insects in a light trap II. The effect of weather conditions on insect activity; and the estimation and forecasting of changes in insect population. *Trans. R. Ent. Soc. Lond.*, 90:227–306.

Williams, D. F., and J. S. Findley. 1979. Sexual size dimorphism in vespertilionid bats. *Am. Midl. Nat.*, 102:113–126.

Williams, T. C., J. M. Williams, and D. R. Griffin. 1966. The homing ability of the neotropical bat *Phyllostomus hastatus*, with evidence for visual orientation. *Anim. Behav.*, 14:468–473.

Wilson, D. E. 1971. Ecology of *Myotis nigricans* on Barro Colorado Island, Panama Canal Zone. *J. Zool. Lond.*, 163:1–13.

Wilson, D. E., and J. S. Findley. 1970. Reproductive cycle of a neotropical insectivorous bat *Myotis nigricans*. *Nature*, 225:1155.

Wilson, D. E., and A. L. Gardner. (eds.). 1980. *Proceedings Fifth International Bat Research Conference*. Lubbock: Texas Tech Univ. Press.

Wilson, E. O. 1975. *Sociobiology, the New Synthesis.* Cambridge: Harvard Univ. Press.

Wimsatt, W. A. 1944. Further studies on the survival of spermatozoa in the female reproductive tract of the bat. *Anat. Rec.,* 88:193–204.

Wimsatt, W. A. (ed.). 1970a. *Biology of Bats,* vol. I. New York: Academic Press.

———(ed.). 1970b. *Biology of Bats,* vol. II. New York: Academic Press.

———1975. Some comparative aspects of implantation. *Biol. Reprod.,* 12:1–40.

———(ed.). 1977. *Biology of Bats.* vol. III. New York: Academic Press.

Wood, F. G., and W. E. Evans. 1980. Adaptiveness and ecology of echolocation in toothed whales. In *Animal Sonar Systems,* ed. R-G. Busnel and J. F. Fish, pp. 381–426. New York: Plenum Press.

Woodsworth, G. C. 1981. Spatial partitioning by two species of sympatric bats, *Myotis californicus* and *Myotis leibii.* M.Sc. Thesis, Department of Biology, Carleton University, Ottawa, Canada.

Yalden, D. W., and P. A. Morris. 1975. *The Lives of Bats.* New York: Demeter Press, Quadrangle, New York Times Book Co.

Zook, J. M., and H. H. Casseday. 1980. Ascending auditory pathways in the brain stem of the bat, *Pteronotus parnellii.* In *Animal Sonar Systems,* ed. R-G. Busnel and J. F. Fish, pp. 1005–1006. New York: NATO Advanced Study Institutes A28, Plenum Press.

INDEX

Italicized numbers indicate references to illustrations

159